T0253563

Satoyama Initiative Thematic Review

Series Editor

Tsunao Watanabe, United Nations University Institute for the Advanced Study of Sustainability (UNU-IAS), Tokyo, Japan

This Open Access book series aims to make timely and targeted contributions for decision-makers and on-the-ground practitioners by producing knowledge concerning "socio-ecological production landscapes and seascapes" (SEPLS) – areas where production activities help maintain biodiversity and ecosystem services in various forms while sustainably supporting the livelihoods and well-being of local communities. Each volume will be designed as a compilation of case studies providing useful knowledge and lessons focusing on a specific theme that is important for SEPLS, accompanied by a synthesis chapter that extracts lessons learned through the case studies to present them for policy-relevant academic discussions. The series is also intended to contribute to efforts being made by researchers and scientists to strengthen the evidence base on social-ecological dynamics and resilience, including those under the Intergovernmental Science-Policy Platform on Biodiversity and Ecosystem Services (IPBES) and the Convention on Biological Diversity (CBD).

The promotion and conservation of SEPLS have been the focus of the Satoyama Initiative, a global effort to realise societies in harmony with nature. In 2010, the International Partnership for the Satoyama Initiative (IPSI) was established to implement the concept of the Satoyama Initiative and promote various activities by enhancing awareness and creating synergies among those working with SEPLS. As a unique platform joined by governmental, intergovernmental, nongovernmental, private-sector, academic, and indigenous peoples' organisations, IPSI has been collecting and sharing the information, lessons, and experiences on SEPLS to accumulate a diverse range of knowledge. Case studies to be published in this series will be solicited from practitioners, researchers, and others working on the ground, who are affiliated with IPSI member organisations and closely involved and familiar with the activities related to SEPLS management.

Maiko Nishi • Suneetha M. Subramanian •
Himangana Gupta

Editors

Biodiversity-Health-Sustainability Nexus in Socio-Ecological Production Landscapes and Seascapes (SEPLS)

Springer

Editors
Maiko Nishi
Biodiversity & Society
United Nations University Institute for the
Advanced Study of Sustainability
Tokyo, Tokyo, Japan

Suneetha M. Subramanian
Biodiversity & Society
United Nations University Institute for the
Advanced Study of Sustainability
Tokyo, Tokyo, Japan

Himangana Gupta
Biodiversity & Society
United Nations University Institute for the
Advanced Study of Sustainability
Tokyo, Tokyo, Japan

ISSN 2731-5169   ISSN 2731-5177   (electronic)
Satoyama Initiative Thematic Review
ISBN 978-981-16-9895-8   ISBN 978-981-16-9893-4   (eBook)
https://doi.org/10.1007/978-981-16-9893-4

Funding Information: United Nations University Institute for the Advanced Study of
Sustainability (UNU-IAS)

© UNU-IAS 2022. This book is an open access publication.
The opinions expressed in this publication are those of the authors/editors and do not necessarily reflect the
views of UNU-IAS, its Board of Directors, or the countries they represent.
Open Access This book is licenced under the terms of the Creative Commons Attribution 3.0 IGO Licence
(http://creativecommons.org/licenses/by/3.0/igo/), which permits use, sharing, adaptation, distribution and
reproduction in any medium or format, as long as you give appropriate credit to UNU-IAS, provide a link
to the Creative Commons licence and indicate if changes were made.
The use of the UNU-IAS name and logo, shall be subject to a separate written licence agreement between
UNU-IAS and the user and is not authorised as part of this CC BY 3.0 IGO licence. Note that the link
provided above includes additional terms and conditions of the licence.
The images or other third party material in this book are included in the book's Creative Commons licence,
unless indicated otherwise in a credit line to the material. If material is not included in the book's Creative
Commons licence and your intended use is not permitted by statutory regulation, or exceeds the permitted
use, you will need to obtain permission directly from the copyright holder.
The use of general descriptive names, registered names, trademarks, service marks, etc. in this publication
does not imply, even in the absence of a specific statement, that such names are exempt from the relevant
protective laws and regulations and therefore free for general use.
The publisher, the authors, and the editors are safe to assume that the advice and information in this book
are believed to be true and accurate at the date of publication. Neither the publisher nor the authors nor the
editors give a warranty, expressed or implied, with respect to the material contained herein or for any
errors or omissions that may have been made. The publisher remains neutral with regard to jurisdictional
claims in published maps and institutional affiliations.

This Springer imprint is published by the registered company Springer Nature Singapore Pte Ltd.
The registered company address is: 152 Beach Road, #21-01/04 Gateway East, Singapore 189721,
Singapore

Review Committee
Nadia Bergamini
Jung-Tai Chao
William Olupot
Fausto O. Sarmiento

Editorial Support
Madoka Yoshino
Kanako Yoshino
Yoshino Nakahara
Vision Bridge, LLC
Evan Namkung
Jennifer Rowton
Ruchika Vemuri

English Proofreading
Susan Yoshimura

Foreword 1

In the context of a global pandemic and climate crisis, the relationship between humans and nature has come under renewed scrutiny. We are currently in the sixth phase of global biodiversity extinction, primarily driven by anthropogenic activities and unsustainable development. This unprecedented biodiversity loss is placing increasing pressures on the planet, and research shows that it is increasing the risk of future pandemics.

As part of our mission to advance global efforts towards sustainability, UNU-IAS has been contributing to the sustainable use of biodiversity at all levels through research, policy engagement, and capacity development initiatives. We have focused on promoting sustainable resource management approaches and positive human-nature interactions in various geographically distinct *socio-ecological production landscapes and seascapes* (SEPLS). Integrated approaches for SEPLS management have the potential to solve various local and regional problems while contributing towards achieving global goals for sustainable development and biodiversity.

This seventh book in the Satoyama Initiative Thematic Review (SITR) series presents a set of case studies that demonstrate critical dynamics in the biodiversity, health, and sustainability nexus in the context of SEPLS. The case studies were selected in light of linkages between biodiversity and health on which understanding is beginning to develop. The studies show how SEPLS management has helped to improve the quality of environments and human lives, with multiple other benefits, making it a valuable tool for enhancing both human and ecosystem health.

The compilation of case studies was produced through the Satoyama Initiative, a global effort to realise societies in harmony with nature. UNU-IAS has worked closely with the Ministry of the Environment of Japan to develop this initiative. We have hosted the Secretariat of the International Partnership for the Satoyama Initiative (IPSI) since its establishment in 2010 at the Tenth Meeting of the Conference of the Parties to the Convention on Biological Diversity (CBD COP 10) in Aichi, Nagoya, Japan. In this role, we coordinate the efforts of partners across the globe towards biodiversity conservation through integrated and holistic landscape and seascape management approaches. IPSI has so far accumulated a diverse range of knowledge and experiences, including a database of 260 case studies submitted by the 283 IPSI member organisations. They include governmental, non-governmental, research, and indigenous peoples' organisations, working collaboratively for better management of SEPLS in various settings around the world.

The SITR publication series was launched in 2015 to share the knowledge of IPSI members and showcase their diverse work across major policy-relevant themes. The previous six volumes each explored a specific set of concerns related to biodiversity in the context of SEPLS. This latest book focuses on the biodiversity-health-sustainability nexus, exploring approaches to build back better from the COVID-19 pandemic.

The volume seeks to provide inspiration and useful knowledge for practitioners, policymakers, and scientists to deepen the understanding of this nexus and advance local solutions to achieve global goals for health, biodiversity, and sustainable development, using traditional and local knowledge. It will also provide a broader contribution to the knowledge base informing key policy processes, including those of the Intergovernmental Science-Policy Platform on Biodiversity and Ecosystem Services (IPBES) and CBD. In particular, the case studies presented here provide relevant and on-the-ground knowledge that will contribute to the ongoing IPBES Nexus Assessment. The volume will help to inform and support the action that is urgently needed to conserve biodiversity and achieve sustainability.

United Nations University   S. Yume Yamaguchi
Institute for the Advanced Study of
Sustainability
Tokyo, Japan

Foreword 2

The Satoyama Initiative seeks to advance the concept of humans living in harmony with nature. This has been the principle on which numerous communities across the globe have traditionally lived, viewing and managing social-ecological systems for their well-being. Recognising the interdependence and reciprocal relationship of nature and societies, human activities in such systems were attuned to ensuring resilience from various disturbances. The intention of the Satoyama Initiative, through the International Partnership for the Satoyama Initiative (IPSI), is to go back to the basics and support the rejuvenation and strengthening of such systems (termed socio-ecological production landscapes and seascapes, SEPLS) that involve a mosaic of ecosystems that are managed in diverse ways to provide multiple benefits to people. Promoting such integrated landscape and seascape management has been identified as a key solution to environmental issues ranging from biodiversity loss and ecosystem degradation to climate regulation by policy forums including the Convention on Biological Diversity (CBD) and the Intergovernmental Science-Policy Platform on Biodiversity and Ecosystem Services (IPBES).

In this context, the Initiative in collaboration with IPSI partners has been engaged in identifying key lessons from such on-the-ground activities to inform various sustainability-related policy priorities. Each volume of the Satoyama Initiative Thematic Review (SITR) focuses on a specific area of policy interest and invites manuscripts detailing the experience of IPSI partners in addressing the theme. Since the sixth edition of SITR, we have collaborated with Springer in publishing.

The manuscripts are peer reviewed and written in a form that can be easily understood by practitioners, policymakers, and of course academics. This volume focuses on an important topic that has come to the forefront of all our lives since

2020: the nexus between biodiversity, health, and sustainable development. I hope that you find reading the different case studies and the analysis in the synthesis chapter interesting and useful.

United Nations University   Tsunao Watanabe
Institute for the Advanced Study of
Sustainability
Tokyo, Japan

International Partnership for the
Satoyama Initiative, Tokyo, Japan

Preface

The Satoyama Initiative is "a global effort to realise societies in harmony with nature", started through a joint collaboration between the United Nations University (UNU) and the Ministry of the Environment of Japan. The initiative focuses on the revitalisation and sustainable management of "socio-ecological production landscapes and seascapes" (SEPLS), areas where production activities help maintain biodiversity and ecosystem services in various forms while sustainably supporting the livelihoods and well-being of local communities. In 2010, the International Partnership for the Satoyama Initiative (IPSI) was established to implement the concept of the Satoyama Initiative and promote various activities by enhancing awareness and creating synergies among those working with SEPLS. IPSI provides a unique platform for organisations to exchange views and experiences and to find partners for collaboration. As of April 2022, 283 members have joined the partnership, including governmental, intergovernmental, non-governmental, private-sector, academic, and indigenous peoples' organisations.

The Satoyama Initiative promotes the concept of SEPLS through a threefold approach that argues for connection of land- and seascapes holistically for the management of SEPLS (see Fig. 1). This often means involvement of several sectors at the landscape scale, under which it seeks to (1) consolidate wisdom in securing diverse ecosystem services and values, (2) integrate traditional ecosystem knowledge and modern science, and (3) explore new forms of co-management systems. Furthermore, activities for SEPLS management cover multiple dimensions, such as equity, addressing poverty and deforestation, and incorporation of traditional knowledge for sustainable management practices in primary production processes such as agriculture, fisheries, and forestry (UNU-IAS & IGES 2015).

As one of its core functions, IPSI serves as a knowledge-sharing platform through the collection and sharing of information and experiences on SEPLS, providing a place for discussion among members and beyond. So far 260 case studies have been collected and are shared on the IPSI website, providing a wide range of knowledge covering diverse issues related to SEPLS. Discussions have also been held to further strengthen IPSI's knowledge facilitation functions, with members suggesting that

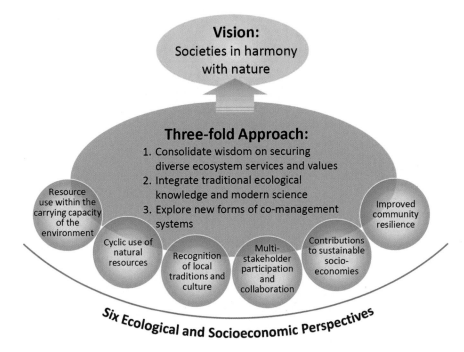

Fig. 1 The conceptual framework of the Satoyama Initiative (source: IPSI Secretariat, 2015)

efforts should be made to produce knowledge on specific issues in SEPLS in order to make more targeted contributions to decision makers and on-the-ground practitioners.

It is in this context that a project to create a publication series titled the "Satoyama Initiative Thematic Review" (SITR)[1] was initiated in 2015. The Thematic Review was developed as a compilation of case studies providing useful knowledge and lessons focusing on a specific theme that is important for SEPLS. The overall aim of the Thematic Review is to collect experiences and relevant knowledge, especially from practitioners working on the ground, considering their usefulness in providing concrete and practical knowledge and information as well as their potential to contribute to policy recommendations. Each volume is also accompanied by a synthesis chapter, which extracts lessons learned through the case studies, presenting them for policy-relevant academic discussions. This series also contributes to efforts being made by researchers to strengthen the evidence base for policymaking concerning social-ecological dynamics and resilience, including those under the Intergovernmental Science-Policy Platform on Biodiversity and Ecosystem Services (IPBES) and the Convention on Biological Diversity (CBD).

---

[1] The previous volumes of the SITR series are available at https://satoyama-initiative.org/featured_activities/sitr/.

Six volumes of the SITR series have been published on an annual basis since 2015. The first volume with the theme "Enhancing knowledge for better management of SEPLS" focused on ways to collect, exchange, refine, and make use of information and knowledge for better management of SEPLS. The second volume's theme was "Mainstreaming concepts and approaches of SEPLS into policy and decision-making", covering topics including advocacy, multi-stakeholder engagement, institutional coordination, tools, and information useful for policymakers and stakeholders. The third volume, titled "Sustainable livelihoods in SEPLS", identified drivers linked to sustainable livelihoods that are crucial to ensure human well-being and foster sustainable natural resource use. The fourth volume, "Sustainable use of biodiversity in SEPLS and its contribution to effective area-based conservation", looked at how effective management of SEPLS, including areas inside and outside of designated protected areas, can achieve benefits for both biodiversity conservation and human livelihoods. The fifth volume, "Understanding the multiple values associated with sustainable use in SEPLS", examined intrinsic, instrumental, and relational values provided and maintained through SEPLS management. The sixth volume, titled "Fostering transformative change for sustainability in the context SEPLS", explored how SEPLS management relates to the idea of transformative change to further the discussion on sustainable transitions. It was the first in the series published by Springer to reach out to a broader range of readers, while the earlier volumes were in-house publications of UNU-IAS.

Building on the past series, SITR has become a Springer book series from this volume in an aim to enhance consistent and coherent contributions to science-policy-practice interfaces while maintaining the publications' impact and reach to a wide audience. Furthermore, the review committee is the first established for this volume to engage in the review process for publication. By inviting experienced and knowledgeable experts in SEPLS management from the IPSI community, this expert group has helped to ensure credibility and reinforce the quality of the contents and at the same time to facilitate sharing of expertise and collaboration among IPSI members.

This publication was developed through a multistage process, including both peer review and discussion among the authors and reviewers at a workshop. Authors had several opportunities to receive feedback, which helped them to improve their manuscripts in substance, quality, and relevance. First, each manuscript received comments from the editorial team and the review committee relating primarily to its contributions to the theme of the volume. Peer review was then conducted by authors of other chapters. Each author received feedback from two other authors who were requested to comment on whether the manuscript was easy to understand and informative, addressed key questions of the volume's focus, and provided useful lessons. The aforementioned workshop was then held virtually on 28–30 June 2021 to enable the exchange of feedback between authors and reviewers. The basic ideas contained in the synthesis of the concluding chapter were developed from presentations and discussions during the workshop, and the chapter was made available for review by authors and reviewers before finalisation.

The above process offers an opportunity for authors from both academic and non-academic organisations to contribute to generating knowledge in an accessible and interactive way, as well as to provide high-quality papers written in simple language for academics and a broader audience alike. It is our hope that this publication will be useful in providing information and insights to practitioners, researchers, and policymakers on the importance of long-term collaborative management of SEPLS for minimising negative trade-offs and creating and enhancing positive synergies between health, biodiversity, and sustainable development to facilitate transformative change towards a sustainable world. This, we hope, will prompt policymaking that strengthens such integrated and holistic management approaches.

We would like to thank all the authors who contributed their case studies. We also appreciate the continued commitment and support for the thorough review process by the four experts in the review committee: Nadia Bergamini, Jung-Tai Chao, William Olupot, and Fausto O. Sarmiento. We are also grateful to the other participants in the case study workshop who provided insightful remarks and valuable inputs into the discussions. These individuals include Alexandros Gasparatos, Hana Matsuzaki, and Eiji Tanaka. We also thank the colleagues of the UNU-IAS and Vision Bridge who were supportive and instrumental in organising the workshop and facilitating the publication process: William Dunbar, Michiko Hashimoto, Yuki Kakuta, Mariko Kinoshita, Michael Lin, Vandana Nagaraji, Yoshino Nakahara, Evan Namkung, Jennifer Rowton, Rizen Tamrakar, Nicholas Turner, Ruchika Vemuri, Makiko Yanagiya, Tomoyo Yanase, Kanako Yoshino, and Madoka Yoshino. Our gratitude extends to the UNU administration, especially Francesco Foghetti and Florence Lo, and to colleagues from the UN Geospatial Information Section for all the instrumental support through the publication process. Furthermore, we acknowledge Susan Yoshimura who skilfully proofread the manuscripts.

Publication of this volume as a new Springer book would not have been possible without the helpful guidance of Mei Hann Lee and Momoko Asawa from Springer and the institutional support and leadership of Shinobu Yume Yamaguchi and Tsunao Watanabe from the UNU-IAS. Our grateful thanks are also due to the Ministry of the Environment, Japan, for supporting the activities of IPSI and its secretariat hosted by the UNU-IAS.

Tokyo, Japan   Maiko Nishi
                                                                      Suneetha M. Subramanian
                                                                            Himangana Gupta

References

IPSI Secretariat. (2015). *IPSI: The International Partnership for the Satoyama Initiative*, viewed 31 October 2021. Retrieved from https://satoyama-initiative.org/wp-content/uploads/2015/11/20151007_ID-PDF_UNU-DL-flyer-EN-with-new-diargam.pdf.

UNU-IAS & IGES (Eds.) (2015). *Enhancing knowledge for better management of socio-ecological production landscapes and seascapes (SEPLS).* Satoyama initiative thematic review, vol. 1. United Nations University Institute for the Advanced Study of Sustainability.

Contents

Editors and Authors

About the Editors

Maiko Nishi Research Fellow at the United Nations University Institute for the Advanced Study of Sustainability. Her research interests include social-ecological system governance, regional planning, and agricultural land policy. PhD in Urban Planning from Columbia University.

Suneetha M. Subramanian Visiting Fellow at the United Nations University Institute for the Advanced Study of Sustainability. Her research interests include biodiversity and human well-being with a focus on equity, traditional knowledge, community well-being, and socio-ecological resilience.

Himangana Gupta Visiting Research Fellow at the United Nations University Institute for the Advanced Study of Sustainability (UNU-IAS), and former JSPS-UNU Postdoctoral Fellow at UNU-IAS and the University of Tokyo. She has worked on climate policy, forestry, biodiversity, and women.

About the Authors

Hennriette Adolf She completed a 6-month internship at the Institute of Interdisciplinary Mountain Research of the Austrian Academy of Sciences in Innsbruck as part of her master's thesis on anthropogenic vegetational changes in the Alps.

Priyanie H. Amerasinghe Scientist Emeritus at the International Water Management Institute, Sri Lanka. She is an eminent zoologist and public health specialist by training with 35 years of experience in environmental research and community health.

Mari Arimitsu Works for the Japan International Cooperation Agency (JICA) focusing on gender equality and gender-based violence. From 2017 to 2021, she worked with government officials and local farmers in Kampong Cham Province, Cambodia, promoting sustainable agriculture.

Julia Atayi Associate Scientist and Project Manager at Conservation Alliance International, Ghana. She is currently coordinating the implementation of a project that aims to enhance communities' adaptive capacity within forest production landscapes.

Oliver Bender Head of the working group "Man and Environment, Settlements" at the Institute of Interdisciplinary Mountain Research of the Austrian Academy of Sciences in Innsbruck.

María Fernanda Cepeda-González Works as a consultant for national and international NGOs and research centres. For 25 years she has focused her career on natural resource management and conservation.

Nicolás Chan-Chuc BSc in Linguistics and Mayan Culture from Eastern University in Yucatán. He is a Mayan-Spanish translator and interpreter and works as a Research Assistant at the Research Center in Geospatial Information Sciences (CentroGeo).

Dipayan Dey Chair of the South Asian Forum for Environment. He is an environmental biotechnologist with expertise in restoration ecology as well as 25 years of comprehensive experience in community-based initiatives in South Asia.

Andrea Fischer Head of the working group "Man and Environment, High Mountains" at the Institute of Interdisciplinary Mountain Research of the Austrian Academy of Sciences in Innsbruck.

Peou Hang Director General of APSARA. He has a PhD in Hydrology and is passionate about rehabilitation of the extensive hydraulic system network within Angkor Park and Khmer ancestors' technology in water management.

Chris Jacobson Adjunct Associate Professor at the University of Queensland. She works as a consultant specialising in adaptation, resilience, and agricultural development in the Asia-Pacific region. She has published more than 40 peer-reviewed articles, book chapters, and books.

Lilian Juárez-Téllez PhD in Ecology and Natural Resource Management from INECOL. She is currently collaborating in CentroGeo Yucatán in territorial studies of complex socioecological systems such as Mayan milpa.

Paulina G. Karimova Master's in Environmental Science, PhD candidate and Research Assistant at the Landscape Conservation and Community Participation Laboratory, National Dong Hwa University, Taiwan. Her research interests include integrated landscape and seascape approaches, adaptive co-management, and community development.

N. Anil Kumar Senior Director at M S Swaminathan Research Foundation, Kerala, India. He has over three decades' professional experience in conservation and sustainable management of genetic resources.

Kuang-Chung Lee PhD in Geography, Professor and Supervisor of the Landscape Conservation and Community Participation Laboratory, National Dong Hwa University, Taiwan. His research focuses on community participation, natural and cultural heritage conservation, collaborative governance of protected areas, and SEPLS.

Andrés López-Rosada Agricultural engineer who majored in sustainable development, ethnodevelopment, and traditional production of indigenous communities. Project Coordinator in the Association of Indigenous Councils of Valle del Cauca-Pacific Region (ACIVA-RP).

Patrick Maundu Ethnobotanist at the National Museums of Kenya and an Honorary Research Fellow at the Alliance of Bioversity International and CIAT. His interest lies in the use of biodiversity by local communities.

María Elena Méndez-López PhD in Environmental Science from the Autonomous University of Barcelona. Her current research interests are traditional agriculture and participation in conservation strategies in rural areas.

Diana M. Mendoza-Salazar Bachelor's in History, MSc in Anthropology. Professor and Researcher at Universidad del Valle. Her lines of research are ethnohistory, anthropology, environmental history, and cultural studies.

Machito Mihara Founder and President of the Institute of Environmental Rehabilitation and Conservation, Japan. He has initiated several projects to promote sustainable agriculture and environmental conservation in Southeast Asian countries.

Yasuyuki Morimoto Associate Scientist at the Alliance of Bioversity International and CIAT in Kenya. He works on projects focused on supporting local communities to adapt to change and create new opportunities for development using crop diversity.

Christian Nielsen Executive Director of the Live & Learn Network with 25 years' experience in international climate change leadership and resilience development.

Paa Kofi Osei-Owusu Social Scientist and Researcher at Conservation Alliance International, Ghana. He is conducting a number of studies to establish factors that influence rural communities' adoption of production technologies within different ecosystem types.

Yaw Osei-Owusu Senior Director and Lead Scientist at Conservation Alliance International, Ghana. He supports the design and development of research-related activities within the organisation.

Raymond Owusu-Achiaw Scientist and Researcher at Conservation Alliance International, Ghana. He is currently conducting a number of studies on the effects of pesticides on the resilience and productive capacity of different ecosystem types.

Md. Shah Paran Researcher at the *Unnayan Onneshan*, Dhaka, Bangladesh. Bachelor's in Development Studies and Master's (major in development economics) from the University of Dhaka.

Rosa Martha Peralta-Blanco Master's degree in Geography from the National Autonomous University of Mexico. She is working as a GIS Expert in the National Geointelligence Laboratory of CentroGeo in Yucatán, Mexico.

Sébastien Proust National Coordinator for the Small Grant Programme in Mexico and works with UNDP. Previously, he dedicated 12 years to working in local and international NGOs on local solutions to address deforestation, clean energy, and climate change adaptation.

Sebastian Orjuela-Salazar Executive Director of Corporación Ambiental y Forestal del Pacifico (CORFOPAL), who majored in planning and declaration of protected areas.

Andrés Quintero-Ángel PhD candidate in Environmental Sciences and Scientific and Research Director of Corporación Ambiental y Forestal del Pacifico (CORFOPAL), who majored in conservation and use of biodiversity with ethnic communities.

David Quintero-Ángel PhD candidate in Agroecology. Bachelor's in Sociology, MBA. Researcher at Corporación Ambiental y Forestal del Pacifico (CORFOPAL). His lines of research are food systems, food diversity, and agroecology.

Mauricio Quintero-Ángel PhD in Environmental Sciences and Associate Professor at Universidad del Valle. Interested in research on social-ecological systems, landscape planning, and rural development.

R. Antonio Riveros-Cañas MSc in Rural Agro-industry, Territorial Development and Tourism from the Autonomous University of Mexico State (UAEMex). Works as a consultant specialist in Rural Agro-industry and Territory at the Inter-American Institute for Cooperation on Agriculture (IICA-Mexico).

Sara Catalina Rodríguez-Díaz Researcher at Corporación Ambiental y Forestal del Pacifico (CORFOPAL) with a Bachelor's in Biology.

Karla Juliana Rodríguez-Robayo PhD in Natural Resource Economics and Sustainable Development from the Universidad Nacional Autónoma de México. In recent years, her research has been focused on the analysis of socioecological systems and socio-environmental conflict.

Jeeranuch Sakkhamduang Works as a Program Manager at the Institute of Environmental Rehabilitation and Conservation to promote sustainable agriculture and reforestation in Cambodia. Currently involved in preparing the national REDD+ strategy for Thailand.

Oscar G. Sánchez-Siordia Director of CentroGeo, Yucatán, and has a PhD in Information Technologies and Computer Systems. His research focuses on Geospatial Data Science.

Andrea A. Serrano-Ysunza Monitoring and Evaluation Specialist at the Small Grant Programme in Mexico. She studied biology and holds a Master's degree in Natural Resources and Rural Development.

V. V. Sivan Senior Scientist at the M S Swaminathan Research Foundation, Kerala, India. He is interested in promoting biodiversity conservation and sustainable utilisation in partnership with local communities.

Jady Smith Landscape Director for Live & Learn Environmental Education. He uses his understanding of environment and communication to strategically support stakeholders by promoting resilience-related initiatives such as socioecological production landscapes.

Socheath Sou Director of Live & Learn Environmental Education, Cambodia. He is passionate about pro-poor innovations in energy, environmental education, natural resource management, climate change, sanitation, and agriculture.

Mariana Rivera-de Velasco BSc in Biology from the Metropolitan Autonomous University (2018). She is a Research Assistant in the Socioecological Systems Area of the Research Center in Geospatial Information Sciences (CentroGeo).

P. Vipindas Development Associate at M S Swaminathan Research Foundation, Kerala, India. He works on projects focused to improve food and nutrition security of Adivasi (Indigenous) communities.

Shao-Yu Yan Master's in Environmental Education, works as a Research Assistant at the Landscape Conservation and Community Participation Laboratory, National Dong Hwa University, Taiwan.

Chapter 1
Introduction

Maiko Nishi, Suneetha M. Subramanian, and Himangana Gupta

Abstract This chapter provides a context for discussing the relevance of socio-ecological production landscapes and seascapes (SEPLS) to the nexus between biodiversity, health, and sustainable development. It begins with an introduction to the idea of a nexus approach to landscape and seascape management, which can help minimise trade-offs and create synergies among different sectors and various global goals for sustainability. With a view to the multiple benefits derived from SEPLS, which extend beyond biodiversity conservation to human and ecosystem health, the chapter then explores how SEPLS management on the ground can contribute to more sustainable management of natural resources, achievement of global targets for biodiversity and sustainable development, and good health for all. Finally, it describes the scope, objectives, and structure of the book, including an overview of the case studies compiled in the subsequent chapters.

Keywords Socio-ecological production landscapes and seascapes · Nexus · Interlinkages · Biodiversity · Health · Sustainability · One Health approach · Case studies · Science-policy-practice interface

1 Biodiversity-Health-Sustainability Nexus

The COVID-19 outbreak, officially declared as a global pandemic on 11 March 2020, has demonstrated the cascading effects of complex human-nature interactions on human health and well-being. Anthropogenic ecosystem changes, including deforestation, agricultural intensification, wildlife exploitation, mining, and infrastructure development, have created a "perfect storm" for the spillover of zoonotic diseases like COVID-19 (Settele et al., 2020). The continued expansion of human activities—including human encroachment into biodiverse habitats, international

M. Nishi (✉) · S. M. Subramanian · H. Gupta
United Nations University Institute for the Advanced Study of Sustainability (UNU-IAS),
Tokyo, Japan
e-mail: nishi@unu.edu

© The Author(s) 2022   1
M. Nishi et al. (eds.), *Biodiversity-Health-Sustainability Nexus in Socio-Ecological Production Landscapes and Seascapes (SEPLS)*, Satoyama Initiative Thematic Review, https://doi.org/10.1007/978-981-16-9893-4_1

movement of people and goods, and unregulated rise in wildlife trade—has significantly increased the risk of zoonotic infections, impacting human lives, economies, and well-being. Pandemics are rather rare compared to small-scale outbreaks but are becoming more frequent due to a ceaseless progress in these underlying events (IPBES, 2020). This will likely lead to an exponential increase in the total cost associated with pandemics, including disease treatments, deaths, and socio-economic impacts (Allen et al., 2017; Berry et al., 2018).

At the same time, nature supports human life and contributes to good health and well-being in numerous ways. "Health" is attributed to not only pathogenic (i.e. disease causing) factors but also factors promoting overall well-being, and is defined as "a state of complete physical, mental, and social well-being and not merely the absence of disease or infirmity" (WHO, 2020; Marselle et al., 2021). Human well-being consists of tangible and intangible elements, including physical well-being (i.e. the quality and performance of bodily functioning), mental well-being (i.e. the psychological, cognitive, and emotional quality of a person's life), and social well-being (i.e. being well connected to others in a local and wider social community) (Linton et al., 2016). Nature offers a wide range of contributions to people, fundamentally securing basic human needs and generally serving as a basis for attaining good quality of life and ensuring human health and well-being. These contributions include food and energy security, access to clean air and water, opportunities for recreation and relaxation, and sense of place (IPBES, 2019b). Biodiversity, comprising the variability among living organisms from all sources and the ecological complexes, underpins these contributions that are essential for human health and well-being (Marselle et al., 2021).

Nature's contributions to people, however, are not always distributed equally to benefit all segments of society, while the costs and burdens associated with their production and use are often borne disproportionally by different groups of people (IPBES, 2019b). Ecosystem changes may benefit some populations but at the expense of others, especially the most vulnerable, signalling crucial trade-offs that need to be managed to pursue sustainable development. The COVID-19 pandemic has exposed this complicated challenge in our society. While all citizens are affected by infectious risks, some evidence shows that vulnerable groups—including not only the elderly and those with ill health and comorbidities, but also people of lower socio-economic status and ethnic minority—are hit harder by the environmental and other related stressors caused by the pandemic (Gaynor & Wilson, 2020; Tavares & Betti, 2021). A subset of factors (e.g. overcrowded accommodation, unstable work conditions and incomes, limited access to healthcare services, non-communicable diseases arising from poverty) conjointly makes those socially deprived and economically disadvantaged more vulnerable to the COVID-19 pandemic, often exacerbating existing inequalities (Amerio et al., 2020; EEA, 2020; Patel et al., 2020).

Rather than negative trade-offs, positive synergies for human health and well-being also exist as we can see, for instance, in sustainable farming practices that improve soil quality, agricultural productivity, and ecosystem functions such as carbon sequestration. Yet, sustainability cannot be achieved without carefully handling trade-offs. This is even clearer today given that many of nature's contributions are declining and cannot be replaced by any current alternatives (IPBES, 2019b).

To better manage trade-offs and create and strengthen positive synergies, the idea of "nexus" has evolved. The term "nexus", originated in the Latin verb *nectare* (meaning "to connect"), has been used in the published literature since the early nineteenth century in various disciplines (e.g. philosophy, governance, cell biology, economics) to trace, describe, and characterise complex interlinkages between multiple objects (Scott et al., 2015; De Laurentiis et al., 2016; Liu et al., 2018). This term was first used in the realm of natural resource use under the Food-Energy Nexus Programme of the United Nations University (UNU) (1983–1988), which aimed to develop an analytical framework and planning methodology for integrated solutions to food and energy scarcity through promoting South-South cooperation in the fields studied (Sachs & Silk, 1990; Scott et al., 2015). Concurrently, the water-resource dimension of the nexus between energy and agriculture started to gain recognition in the western United States in the mid-1980s, considering energy and environmental needs for water, agricultural-irrigation, and urban-industrial demands (Scott et al., 2015). Finally, nexus thinking has been increasingly applied to the study of interconnections between food, water, and energy, often called WEF or FEW nexus, mostly in the climate change context, but sometimes additionally with biodiversity conservation and human health (Albrecht et al., 2018; IPBES, 2019a).

The Global Assessment Report of the Intergovernmental Science-Policy Platform on Biodiversity and Ecosystem Services (IPBES) defines the concept of nexus as "a perspective which emphasizes the inter-relatedness and interdependencies of ecosystem components and human uses, and their dynamics and fluxes across spatial scales and between compartments," adapting the definition posed by UNU Institute for Integrated Management of Material Fluxes and of Resources (UNU-FLORES) (IPBES, 2019a, p. 1047). This definition implies both a phenomenological perspective to capture the interactions among different subsystems (or sectors) within the nexus system and an analytic perspective to examine the links between the nexus nodes (e.g. water and energy), into either of which many varying definitions can be categorised (Zhang et al., 2018). The nexus perspective allows for better understanding of the functioning, productivity, and management of a complex system which involves trade-offs, and facilitation and amplification between the different components (IPBES, 2019a). This approach can thus help detect and minimise negative trade-offs, uncover positive synergies, and unveil unexpected consequences arising from these combined effects (Liu et al., 2018). This effort finally helps enhance integrated planning, governance, and management, and in particular helps identify and pursue pathways to achieve global goals and targets for sustainable development, many of which are interconnected (Nilsson et al., 2016; Weitz et al., 2018).

With this understanding, IPBES (2019a) drew on a nexus approach to analyse interactions between multiple sectors and objectives and identify key elements of sustainable pathways. For the sake of feasibility and comprehensibility for the analysis in the complex context, it applied the approach via six complementary

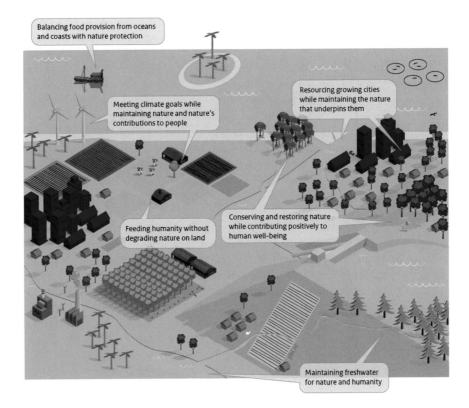

Balancing food provision from oceans
and coasts with nature protection

Meeting climate goals while
maintaining nature and nature's
contributions to people

Resourcing growing cities
while maintaining the nature
that underpins them

Feeding humanity without
degrading nature on land

Conserving and restoring nature
while contributing positively to
human well-being

Maintaining freshwater
for nature and humanity

Fig. 1.1 The six interconnected foci of the nexus analysis in the IPBES Global Assessment (source: Chan et al., 2019). The six foci for the nexus analysis reflect key challenges in conserving nature and nature's contributions to people while achieving the SDGs given both trade-offs and synergies. These foci include (1) feeding humanity without degrading terrestrial natural resources; (2) meeting climate goals without incurring massive land-use change and biodiversity loss; (3) conserving and restoring nature on land while contributing positively to human well-being; (4) maintaining freshwater for nature and humanity; (5) balancing food provision from oceans and coasts with biodiversity protection; and (6) resourcing growing cities while maintaining the ecosystems and biodiversity that underpin them

foci (or lenses) to examine specific links between terrestrial, marine, and freshwater social-ecological systems in consideration of linkages to other entities (Fig. 1.1). This analysis revealed trade-offs and synergies in each of the foci as well as common threads in achieving a subset of sustainable development goals (SDGs) simultaneously. Pointing to significantly varying pathways across geographic contexts, the cross-scale nexus analysis highlighted the significance in integrating local and regional perspectives in global pathways towards sustainability. Despite the diversity of the pathways with different changes needed to achieve global goals at all scales, the analysis also suggested the following common constituents of sustainable pathways associated with seven SDGs relevant to nature (i.e. SDGs 2, 3, 6, 11, 13, 14, and 15): (1) safeguarding remaining natural habitats on land and sea,

(2) undertaking large-scale restoration of degraded habitats, and (3) integrating these activities with development through sustainable planning and management of landscapes and seascapes.

Cross-sectoral cooperation and planning have been increasingly called for to attain global environmental and societal goals, but effective policy integration, for instance across multilateral environmental agreements, is yet to be achieved (Azizi et al., 2019; IPBES, 2019a; van den Heuvel et al., 2020). Also, the nexus approach is still in its infancy for application and implementation, despite its great potential to promote cooperation, coordination, and policy coherence among different sectors (Liu et al., 2018). In this context, IPBES launched a new 3-year thematic assessment of "the interlinkages among biodiversity, water, food, and health" (the so-called nexus assessment) at its eighth plenary session in 2021 to be conducted between 2022 and 2024.[1] This assessment aims to advance understanding on the interlinkages among biodiversity, climate change, adaptation, and mitigation, including relevant aspects of energy, water, food, and health. It will also consider holistic approaches based on different knowledge systems to achieve global goals such as the 2050 Vision for Biodiversity and the 2030 Agenda for Sustainable Development, including those to attain good health for all.

In addition, IPBES organised a virtual workshop on the links between biodiversity and pandemics on 27–31 July 2020 to strengthen the knowledge base on links between biodiversity and current and future pandemics in response to the extraordinary situation caused by the COVID-19 pandemic. The report from this workshop (IPBES, 2020) warns of a daunting future, where without preventive measures, pandemics will emerge more frequently and spread more rapidly, resulting in more human deaths and more devastating socio-economic impacts. Pointing to human activities as the fundamental driver of the emergence of pandemics, the report also highlights the interconnectedness of the world community and the increasing threats to health and well-being arising from global inequality. The report, however, also says that escaping the era of pandemics is still possible, but requires transformative change to shift from the current reactive approach to a preventive one to prepare for future pandemics. Finally, it offers several policy options to foster transformative change and move towards preventing pandemics based on a "One Health" approach (i.e. an approach that integrates human health, animal health, and environmental sectors).

The One Health approach builds on the idea that human health, ecosystem health, and animal health are all interrelated, and that it is imperative to encourage mechanisms that ensure coordination and collaboration among the relevant sectors to strengthen the links between them (WHO, FAO,, & OIE, 2019). From a heuristic view, we understand that the One Health approach is highly compatible with the

---

[1]The IPBES Plenary at its eighth session in June 2021 approved the undertaking of the nexus assessment as outlined in the scoping report set out in annex 1 to its decision IPBES-8/1 for consideration by the Plenary at its eleventh session. The scoping report is available at: https://ipbes. net/sites/default/files/2021-07/20210719_scoping_report_for_the_nexus_assessment.pdf.

approach to managing socio-ecological production landscapes and seascapes (SEPLS). As discussed in the following section, SEPLS have been shaped and managed through strongly interlinked sets of traditional practices and production activities that have been adapted and transformed to maintain and improve the well-being of communities while absorbing shocks to the system, suggesting higher levels of resilience (Bergamini et al., 2013). To evade the pandemic era and achieve global goals for biodiversity and sustainability, strategies to build and strengthen resilience against systemic threats will be increasingly important and relevant in the coming decades. In this regard, the nexus analysis of SEPLS management should offer ideas and available means to devise such strategies.

2   Socio-Ecological Production Landscapes and Seascapes and Nexus Approaches

SEPLS are multifunctional and utilitarian conceptualisations of landscape and sea-scape use. Here the proximate population has recognised its reliance on the SEPLS for various material and intangible benefits and clearly acknowledged the interdependence of social and ecological components of a social-ecological system that is consequently factored into related decisions on the management of the SEPLS. Furthermore, the well-being of the population dependent on a SEPLS is entrenched in ensuring that multiple needs, such as securities of food, health, and energy, identity, culture, and ecological integrity, are met at the same time (Bergamini et al., 2013). This indicates that nexus approaches across different "sectoral priorities" need to be practiced in SEPLS contexts to simultaneously guarantee that these different needs are realised and are aligned during the implementation of relevant activities (Sarmiento & Frolich, 2020). It therefore is a natural extension in the reasoning that concepts such as "Community Health"[2] (Unnikrishnan & Suneetha, 2012) or "One Health" are understood and practiced de facto in local communities as the concept of health encompasses multiple dimensions from access to resources, and well-functioning ecosystems, to food and nutritional security, access to medical resources, and cultural practices, among others.

That said, contexts in which SEPLS operate have changed over time due to changing sociopolitical priorities and other factors. Consequently, decisions related to multiple activities in a SEPLS could be said to have become more sectoral or compartmentalised. This has brought on challenges to ensure the sustenance of endogenous nexus approaches that once were widely practiced—often requiring

---

[2] Community Health builds on the concept of health held by local and indigenous communities that relates not just to medical services, but involves access to food and nutritional security, access to cultural resources, medicinal resources, access to areas of cultural importance, rights to use and practice, and livelihood security.

purposive rejuvenation efforts, as the case studies in this volume will highlight. Drivers of these changes range from policy pressures (e.g. mono or simple cropping patterns, land-use changes for other development purposes) to changes in demographic profiles and priorities (e.g. in- and outmigration) and natural vulnerabilities (e.g. to floods or other natural calamities) (IPBES, 2019a).

As mentioned above, several global assessments are finding increasing evidence that tackling biodiversity loss and ensuring human well-being thereof are linked closely to effective and integrated management of landscapes and seascapes and strengthening the capacities and resources available to the local communities managing them. The increasingly robust emphasis by policy bodies to ensure biodiversity conservation, health, and sustainable development through calls for adopting coherent policies (such as the draft Global Biodiversity Framework or the One Health implementation mechanisms being set up in different countries) serves as a timely opportunity to refocus on the principles underlying SEPLS management—as context-dependent, multi-stakeholder, and multisectoral approaches that are designed to derive multiple benefits. Proof of this concept can be found in the experiences of the members of the International Partnership for the Satoyama Initiative (IPSI). Experiential knowledge and practical lessons in dealing with the nexus between the sectors can be found in the case studies, which can be taken advantage of as approaches to landscape and seascape management.

3 Objectives and Structure of the Book

This book focuses on the biodiversity-health-sustainability nexus in the context of the management and multiple benefits of SEPLS. The primary aim is to provide insights on how SEPLS management on the ground can contribute to enhanced ecosystem and human health, sustainable management of natural resources, and achievement of global targets for biodiversity and sustainable development.

To explicitly showcase the dynamics of the biodiversity-health-sustainability nexus in SEPLS, this volume brings together case studies on SEPLS management from different regions around the world, which delve into the relevance of SEPLS to various aspects of sustainability in the context of ecosystem and human health. The case studies highlight the roles, attitudes, and actions of those responsible for management, including smallholders, indigenous peoples, local communities, and other stakeholders, in conserving biodiversity—including agro-biodiversity—while ensuring the health of SEPLS and dependent communities. For example, they pertain to efforts that enhance ecosystem and human health, such as pesticide-free food production, high nutrition, good water quality, and sustainable tourism. Additionally, most case studies touch upon the impact of the COVID-19 pandemic on the livelihoods of ecosystem-dependent local communities and related changes in ecosystem health. In particular, the case studies address the following questions:

- What multiple benefits derived from SEPLS management have helped to ensure and enhance aspects of human health, and how?
- What are the trade-offs and synergies between efforts to attain ecosystem health, human health, and sustainability in managing SEPLS?
- How can we measure the effectiveness of SEPLS management in securing and improving ecosystem and human health, as well as sustainability?
- What are the challenges and opportunities in managing SEPLS to achieve biocultural diversity conservation and sustainable development while ensuring and enhancing ecosystem and human health?

The case studies commonly address the above key questions to elucidate the relevance of SEPLS management for long-term sustainability, ecosystem health, and human well-being. Chapters 2–12 present 11 case studies encompassing different types of ecosystems around the world, including six from Asia, two from Latin America, two from Africa, and one from Europe (Fig. 1.2). The case studies are divided into four broad thematic areas: (1) local and indigenous conceptualisation of health and well-being; (2) wider landscapes and seascapes and resilience; (3) water, tourism, and recreation; and (4) food and farming. Table 1.1 shows specific SEPLS types and key challenges faced by them in the case studies under each of the thematic areas.

Most of the cases primarily focus on the nexus at the local level, involving local communities and their health (Chaps. 2–5, 9–12) or ecosystem health and related human well-being in the context of recreation and tourism (Chaps. 6–8). To represent the nexus dynamics, each case study illustrates a unique linkage type (e.g. food–health, water management–well-being, rights–well-being, conservation–human well-being) in their SEPLS. The various interconnected challenges (e.g. pollution, climate change, deforestation, biodiversity loss, degradation, COVID-19 pandemic) are addressed using a combination of environmental (e.g. sustainable agricultural practices, enhanced water and ecosystem management), economic (e.g. economic diversification and opportunities, livelihood improvement), and community-oriented (e.g. promotion of traditional knowledge, capacity development, local documentation, improvement in health and well-being of local people) solutions. Importantly, many of the cases exemplify initiatives not only to address immediate problems, but also to collectively identify long-term solutions and ensure continuous delivery of multiple benefits from SEPLS, while maintaining and enhancing ecosystem and human health.

To conclude with the key findings from these case studies, Chap. 13 distils the relevant messages to offer implications for science, policy, and practice as well as their interfaces in better managing the biodiversity-health-sustainability nexus in the context of SEPLS. The synthesis of the case studies' findings offers relevant insights into the local-level implementation of nexus approaches and methodologies for monitoring and evaluation, using localised tools and indicators compatible with global ones, and involving multiple disciplines and sectors relevant to nexus approaches.

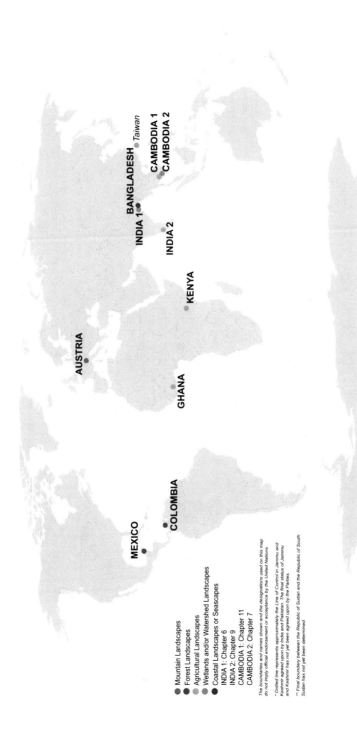

Fig. 1.2 Locations of the case studies (regions, landscapes, and/or seascapes) (map template: Geospatial Information Section, United Nations). Note: Details of the case study locations, including geographic coordinates, are described in each chapter

Table 1.1 Overview of the case studies

Focused thematic areas	Chapter (Country)	SEPLS types	Key interlinkages	Problems and Challenges	Objectives
Local and indigenous conceptualisation of health and well-being	Chapter 2 (Bangladesh)	Forest, inland water, coastal and marine, wetland	Rights, conservation, well-being, health	Mangrove degradation, livelihood insecurity, COVID-19 outbreak	Conceptualise one health approach, enhance livelihood security, and promote customary sustainable practices and traditional knowledge
	Chapter 3 (Colombia)	Forest, traditional agriculture	Biocultural memory, conservation, human well-being, securing territorial rights	Loss of biodiversity, loss of cultural identity, armed conflict, poverty	Recover traditional knowledge and cultural conservation values to enhance quality of health
Wider landscapes and seascapes and resilience	Chapter 4 (Chinese Taipei)	Mountain, forest, agriculture, watershed, coastal, mountains	Eco-agriculture, SEPLS well-being, landscape-seascape connectivity	Dams and river dredging, tourism, loss of native varieties, outmigration	Promote eco-agriculture, weaving of traditional and modern knowledge, and cooperation between multiple stakeholders across different ecosystems at a landscape-seascape scale
	Chapter 5 (Mexico)	Forest, traditional agriculture	Agrochemicals, sustainable landscape management, health, cultural heritage	Pollution, recurrent diseases	Promote sustainable management of ecosystems, enhance landscape resilience, and raise awareness on traditional food production
Water, tourism, and recreation	Chapter 6 (India—peri-urban wetlands)	Peri-urban, wetlands	Bio-rights, health, climate	Pollution, lack of awareness	Create economic opportunities, enrich biodiversity, and improve human health and well-being
	Chapter 7 (Cambodia—Angkorian landscape)	Forest, watershed, agriculture, peri-urban, urban	Water management, forest management, tourism, human well-being	Climate change, deforestation, declined groundwater recharge, increased demand for food and water	Enhance water management, promote economic diversification, preserve cultural heritage, and improve human health and well-being
	Chapter 8 (Austria)	Mountain, forest	Human health, biodiversity	Climate change, tourism	Revitalise mountain livelihoods, and improve lifestyle

	Chapter				
Food and farming	Chapter 9 (India—a biocultural hotspot)	Agriculture, mountain, forest	Food, health	Degradation of SEPLS, loss of biodiversity, climate change, food and nutritional insecurity	Promote local health traditions and local health baskets, and enhance human immunity to infectious diseases
	Chapter 10 (Kenya)	Forest, agriculture, semi-arid land	Food, nutrition, biodiversity, local and traditional knowledge, human health	Underutilised local foods, stigma, loss of cultural heritage, obesity, biased nutrition	Preserve traditional knowledge and cultural practices, create economic and livelihood opportunities, enhance human health and well-being, and ensure biological and cultural diversity
	Chapter 11 (Cambodia—agro-biodiverse landscapes)	Agriculture	Food security, human health	COVID-19 pandemic, food and health insecurity	Promote sustainable agriculture, facilitate capacity building, promote organic fertilisers, and enhance agricultural biodiversity
	Chapter 12 (Ghana)	Forest, agriculture	Pesticides, agriculture, human and ecosystem health	Highly hazardous pesticides, pollution, and health impacts on humans and ecosystems	Phase out highly hazardous pesticides, promote integrated pest management, improve human and ecosystem health, and enhance agricultural cocoa production landscape

References

Albrecht, T. R., Crootof, A., & Scott, C. A. (2018). The water-energy-food nexus: A systematic review of methods for nexus assessment. *Environmental Research Letters, 13*(4), 043002. https://doi.org/10.1088/1748-9326/aaa9c6

Allen, T., Murray, K. A., Zambrana-Torrelio, C., Morse, S. S., Rondinini, C., Di Marco, M., Breit, N., Olival, K. J., & Daszak, P. (2017). Global hotspots and correlates of emerging zoonotic diseases. *Nature Communications, 8*(1), 1124. https://doi.org/10.1038/s41467-017-00923-8

Amerio, A., Aguglia, A., Odone, A., Gianfredi, V., Serafini, G., Signorelli, C., & Amore, M. (2020). Covid-19 pandemic impact on mental health of vulnerable populations. *Acta Biomedica Atenei Parmensis, 91*(9-S), 95–96. https://doi.org/10.23750/abm.v91i9-S.10112

Azizi, D., Biermann, F., & Kim, R. E. (2019). Policy integration for sustainable development through multilateral environmental agreements: An empirical analysis, 2007–2016. *Global Governance: A Review of Multilateralism and International Organizations, 25*(3), 445–475. https://doi.org/10.1163/19426720-02503005

Bergamini, N., Blasiak, R., Eyzaguirre, P., Ichikawa, K., Mijatovic, D., Nakao, F., & Subramanian, S. M. (2013). *Indicators of resilience in socio-ecological production landscapes (SEPLs),* UNU-IAS policy report. United Nations University Institute of Advanced Studies.

Berry, K., Allen, T., Horan, R. D., Shogren, J. F., Finnoff, D. C., & Daszak, P. (2018). The economic case for a pandemic fund. *EcoHealth, 15*(2), 244–258. https://doi.org/10.1007/s10393-018-1338-1

Chan, K. M. A., Agard, J., Liu, J., Aguiar, A. P. D., Armenteras, D., Boedhihartono, A. K., Cheung, W. W. L., Hashimoto, S., Pedraza, G. C. H., Hickler, T., Jetzkowitz, J., Kok, M., Murray-Hudson, M., O'Farrell, P., Satterfield, T., Saysel, A. K., Seppelt, R., Strassburg, B., Xue, D., Selomane, O., Balint, L., & Mohamed, A. (2019). Chapter 5. Pathways towards a sustainable future. In: E. S. Brondíizio, J. Settele, S. Dïaz, H. T. Ngo (eds.), *Global assessment report of the intergovernmental science-policy platform on biodiversity and ecosystem services.* IPBES Secretariat. 108 pages. https://doi.org/10.5281/zenodo.3832099.

De Laurentiis, V., Hunt, D. V. L., & Rogers, C. D. F. (2016). Overcoming food security challenges within an energy/water/food nexus (EWFN) approach. *Sustainability, 8*(1), 95. https://doi.org/10.3390/su8010095

EEA. (2020). *Healthy environment, healthy lives: How the environment influences health and well-being in Europe.* European Environment Agency, viewed 11 September 2021. Retrieved from https://www.eea.europa.eu/publications/healthy-environment-healthy-lives.

Gaynor, T. S., & Wilson, M. E. (2020). Social vulnerability and equity: The disproportionate impact of COVID-19. *Public Administration Review, 80*(5), 832–838. https://doi.org/10.1111/puar.13264

van den Heuvel, L., Blicharska, M., Masia, S., Sušnik, J., & Teutschbein, C. (2020). Ecosystem services in the Swedish water-energy-food-land-climate nexus: Anthropogenic pressures and physical interactions. *Ecosystem Services, 44*, 101141. https://doi.org/10.1016/j.ecoser.2020.101141

IPBES. (2019a). In E. S. Brondízio, J. Settele, S. Díaz, & H. T. Ngo (Eds.), *Global assessment report of the intergovernmental science-policy platform on biodiversity and ecosystem services.* IPBES Secretariat. https://doi.org/10.5281/zenodo.3831674

IPBES. (2019b). In S. Díaz, J. Settele, E. S. Brondízio, H. T. Ngo, M. Guèze, J. Agard, A. Arneth, P. Balvanera, K. A. Brauman, S. H. M. Butchart, K. M. A. Chan, L. A. Garibaldi, K. Ichii, J. Liu, S. M. Subramanian, G. F. Midgley, P. Miloslavich, Z. Molnár, D. Obura, A. Pfaff, S. Polasky, A. Purvis, J. Razzaque, B. Reyers, R. R. Chowdhury, Y. J. Shin, I. J. Visseren-Hamakers, K. J. Willis, & C. N. Zayas (Eds.), *Summary for policymakers of the global assessment report on biodiversity and ecosystem services of the intergovernmental science-policy platform on biodiversity and ecosystem services.* IPBES Secretariat, viewed 14 September 2021. Retrieved from https://ipbes.net/sites/default/files/2020-02/ipbes_global_assessment_report_summary_for_policymakers_en.pdf.

IPBES. (2020). In P. Daszak, J. Amuasi, C. G. das Neves, D. Hayman, T. Kuiken, B. Roche, C. Zambrana-Torrelio, P. Buss, H. Dundarova, Y. Feferholtz, G. Földvári, E. Igbinosa, S. Junglen, Q. Liu, G. Suzan, M. Uhart, C. Wannous, K. Woolaston, P. Mosig Reidl, K. O'Brien, U. Pascual, P. Stoett, H. Li, & H. T. Ngo (Eds.), *Workshop report on biodiversity and pandemics of the intergovernmental platform on biodiversity and ecosystem services.* IPBES Secretariat. https://doi.org/10.5281/ZENODO.4147317

Linton, M.-J., Dieppe, P., & Medina-Lara, A. (2016). Review of 99 self-report measures for assessing well-being in adults: Exploring dimensions of well-being and developments over time. *BMJ Open, 6*(7), e010641. https://doi.org/10.1136/bmjopen-2015-010641

Liu, J., Hull, V., Godfray, C., Tilman, D., Gleick, P. H., Hoff, H., Pahl-Wostl, C., Xu, Z., Chung, M. G., Sun, J., & Li, S. (2018). Nexus approaches to global sustainable development. *Nature Sustainability, 1*(9), 466–476. https://doi.org/10.1038/s41893-018-0135-8

Marselle, M. R., et al. (2021). Pathways linking biodiversity to human health: A conceptual framework. *Environment International, 150*, 106420. https://doi.org/10.1016/j.envint.2021.106420

Nilsson, M., Griggs, D., & Visbeck, M. (2016). Policy: Map the interactions between sustainable development goals. *Nature, 534*(7607), 320–322. https://doi.org/10.1038/534320a

Patel, J. A., Nielsen, F., Badiani, A. A., Assi, S., Unadkat, V. A., Patel, B., Ravindrane, R., & Wardle, H. (2020). Poverty, inequality and COVID-19: The forgotten vulnerable. *Public Health, 183*, 110–111. https://doi.org/10.1016/j.puhe.2020.05.006

Sachs, I., & Silk, D. (1990). *Food and energy—Strategies for sustainable development.* United Nations University Press, viewed 11 September 2021. Retrieved from https://archive.unu.edu/unupress/unupbooks/80757e/80757E00.htm.

Sarmiento, F. O., & Frolich, L. M. (Eds.). (2020). *The Elgar companion to geography, transdisciplinarity and sustainability.* Edward Elgar Publishing. https://doi.org/10.4337/9781786430106

Scott, C., Kurian, M., & Wescoat, J. (2015). The water-energy-food nexus: Enhancing adaptive capacity to complex global challenges. In M. Kurian & R. Ardakanian (Eds.), *Governing the nexus: Water, soil and waste resources considering global change* (pp. 15–38). https://doi.org/10.1007/978-3-319-05747-7_2

Settele, J., Díaz, S., Brondizio, E., & Daszak, P. (2020). *IPBES guest article: COVID-19 stimulus measures must save lives, protect livelihoods, and safeguard nature to reduce the risk of future pandemics*, viewed 11 September 2021. Retrieved from http://ipbes.net/covid19stimulus.

Tavares, F. F., & Betti, G. (2021). The pandemic of poverty, vulnerability, and COVID-19: Evidence from a fuzzy multidimensional analysis of deprivations in Brazil. *World Development, 139*, 105307. https://doi.org/10.1016/j.worlddev.2020.105307

Unnikrishnan, P. M., & Suneetha, M. S. (2012). *Biodiversity, traditional knowledge and community health: Strengthening linkages*, UNU-IAS policy report. UNU-IAS & UNEP.

Weitz, N., Carlsen, H., Nilsson, M., & Skånberg, K. (2018). Towards systemic and contextual priority setting for implementing the 2030 agenda. *Sustainability Science, 13*(2), 531–548. https://doi.org/10.1007/s11625-017-0470-0

WHO, FAO, & OIE. (2019). *Taking a multisectoral, one health approach: A tripartite guide to addressing zoonotic diseases in countries.* World Health Organization.

WHO. (2020). Basic documents: Forty-ninth edition (including amendments adopted up to 31 May 2019). World Health Organization, viewed 14 September 2021. Retrieved from https://apps.who.int/gb/bd/pdf_files/BD_49th-en.pdf.

Zhang, C., Chen, X., Li, Y., Ding, W., & Fu, G. (2018). Water-energy-food nexus: Concepts, questions and methodologies. *Journal of Cleaner Production, 195*, 625–639. https://doi.org/10.1016/j.jclepro.2018.05.194

The opinions expressed in this chapter are those of the author(s) and do not necessarily reflect the views of UNU-IAS, its Board of Directors, or the countries they represent.

Open Access This chapter is licenced under the terms of the Creative Commons Attribution 3.0 IGO Licence (http://creativecommons.org/licenses/by/3.0/igo/), which permits use, sharing, adaptation, distribution and reproduction in any medium or format, as long as you give appropriate credit to UNU-IAS, provide a link to the Creative Commons licence and indicate if changes were made.

The use of the UNU-IAS name and logo, shall be subject to a separate written licence agreement between UNU-IAS and the user and is not authorised as part of this CC BY 3.0 IGO licence. Note that the link provided above includes additional terms and conditions of the licence.

The images or other third party material in this chapter are included in the chapter's Creative Commons licence, unless indicated otherwise in a credit line to the material. If material is not included in the chapter's Creative Commons licence and your intended use is not permitted by statutory regulation or exceeds the permitted use, you will need to obtain permission directly from the copyright holder.

Chapter 2
Human-Nature Cooperation for Well-Being: Community Understanding on One Health Approach in the COVID-19 Era in the Sundarbans

Rashed Al Mahmud Titumir and Md. Shah Paran

Abstract This study attempts to explore the interdependent relationship between humans and nature, and to comprehend the community understanding of the "One Health" approach in the face of the COVID-19 pandemic in the Sundarbans in Bangladesh. It explores challenges in socio-ecological production landscapes and seascapes (SEPLS) management, response of indigenous peoples and local communities (IPLCs), and corresponding outcomes, and also examines factors affecting the ecosystem's balance. It particularly draws on the insights of traditional resource users (TRUs) in a part of the Sundarbans who are wood collectors (*Bawali*), fishermen (*Jele*), honey and wax collectors (*Mouali*), and crab collectors. The study adopts a multiple evidence base (MEB) approach in order to bring in the participatory insights of IPLCs, coupled with scientific knowledge and interdisciplinary heterodox perspectives. Based on the community conceptualisation of the One Health approach, this study demonstrates that the appropriation of nature (conservation, restoration, sustainable use, access, and benefit sharing) instead of expropriation (anthropogenic pressures) can serve as a yardstick to ensure a virtuous cycle in the ecosystem and a harmonious relationship between humans and nature. The study presents a modified One Health framework for the post-2020 period that calls for ensuring rights-oriented universal social entitlements, provision of livelihood security, and promotion of human-nature cooperation underwritten by customary sustainable practices and traditional knowledge in SEPLS management.

Keywords Human-nature cooperation · Well-being · One Health approach · COVID-19 · SEPLS · Sundarbans · Bangladesh

R. A. M. Titumir (✉)
Department of Development Studies, University of Dhaka, Dhaka, Bangladesh

The Unnayan Onneshan, Dhaka, Bangladesh
e-mail: rtitumir@unnayan.org; rt@du.ac.bd

Md. S. Paran
The Unnayan Onneshan, Dhaka, Bangladesh

© The Author(s) 2022   15
M. Nishi et al. (eds.), *Biodiversity-Health-Sustainability Nexus in Socio-Ecological Production Landscapes and Seascapes (SEPLS)*, Satoyama Initiative Thematic Review, https://doi.org/10.1007/978-981-16-9893-4_2

1 Introduction

The study explores the interrelationship between human beings and nature that directly and indirectly affects both, and conjoins participatory research approaches to ascertain community understanding on the "One Health" approach that encompasses human, animal, and ecosystem health together in the COVID-19 era in the Sundarbans. It explores the challenges in socio-ecological production landscapes and seascapes (SEPLS) management and the responses of the traditional resource users (TRUs) and corresponding outcomes. Factors affecting the ecosystem's balance, and thus the well-being of the forest and its people, are also examined. It further examines the state of health and well-being of indigenous peoples and local communities (IPLCs) and TRUs in terms of universal access to health and social security programmes, as well as the state of ecosystem health in terms of biodiversity, anthropogenic pressures, resource management, and sustainable production and consumption in the face of COVID-19. Accordingly, the study categorises the human contribution to nature (biodiverse adaptation to climate change, nature-based community solutions for livelihood diversification, customary sustainable practices based on traditional knowledge) and nature's contribution to human beings in promoting One Health in the Sundarbans.

The Sundarbans, situated at the edge of the Bay of Bengal, is the largest neighbouring single-tract contiguous mangrove ecosystem in the world. It is a unique SEPLS with a composite ecosystem combining forest, marine, coastal, and wetland environments, located in the southwest corner of Bangladesh, between $21°30'$ and $21°39'$ N, $89°01'$ and $89°52'$ E (Fig. 2.1 and Table 2.1). It supports viviparous plant species with 334 species of trees, shrubs, herbs, and epiphytes and around 400 species of wild animals (Behera & Haider, 2012).

A significant number of people maintain their livelihoods by utilising the resources of the forest, and thus the area is a unique hotspot for biodiversity conservation and sustainable use and is identified as a SEPLS. The benefits from the Sundarbans appear in the form of multiple goods and services, which "contribute

(a)   (b)

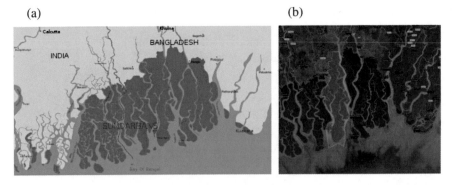

Fig. 2.1 (**a**) Map of the study site (source: Wikimedia Commons Contributor Nirvik12, 2015); (**b**) land cover map of case study site (source: Map data (c) Google, 2021)

Table 2.1 Basic Information of the study area

Country	Bangladesh
Province	n.a.
District	Khulna, Satkhira, and Bagerhat
Size of geographical area (hectare)	607,100
Number of direct beneficiaries (persons)	1300
Number of indirect beneficiaries (persons)	3.5 million
Dominant ethnicity(ies), if appropriate	Bangalee
Size of the case study/project area (hectare)	177,500
Geographic coordinates (latitude, longitude)	21°30′ and 21°39′ N, 89°01′ and 89°52′ E

to making human life both possible and worth living" (Díaz et al., 2006; Millennium Ecosystem Assessment, 2005; Layke et al., 2012; van Oudenhoven et al., 2012). However, the Sundarbans at present is an ecologically vulnerable area due to degradation of biodiversity resources. Over the years, it has experienced major ecological and physiographical changes and is losing its resources due to both human interventions and climatic changes (Titumir & Afrin, 2017). The area of the Bangladesh part of the Sundarbans was 17,000 km^2 in 1776, which has subsequently been reduced to almost half the size (Islam & Gnauck, 2009).

Moreover, the COVID-19 pandemic has posed unprecedented challenges to people's lives and livelihoods in the Sundarbans. For example, around 95% of TRUs in the Sundarbans lost the massive share of their income during the nationwide lockdown (Unnayan Onneshan, 2020c). On top of this, the coast of Bangladesh was struck by supercyclone *Amphan* on 16 May 2020, leaving crops, infrastructure, and coastal protection embankments damaged in 26 coastal districts, which further negatively affected the livelihood options of IPLCs in the Sundarbans (New Age, 2020). Moreover, due to the absence of universal social security programmes in the country, the economic fallout of the majority of the people was exacerbated (Unnayan Onneshan, 2020b). The ongoing health crisis and livelihood insecurity resulting from COVID-19 are marginalising the forest people and also creating a metabolic rift—a disruption in the ecosystem balance caused by a break in the producer, consumer, and decomposer cycle in the ecosystem. This imbalance in the ecosystem cycle has led to the increased ill-being of the forest and its people.

2 Methodology

The case study was conducted adopting a multiple evidence base (MEB) approach. It draws on indigenous and ecosystem-based solutions in SEPLS management, utilising the indigenous and local knowledge (ILK) collected from two cooperatives—*Koyra Bonojibi Bohumukhi Unnayan Samity* and *Munda Adivasi Bonojibi Bohumukhi Unnayan Samity*—in the Koyra Upazila of Khulna District in the southwestern region of Bangladesh, a part of the Sundarbans SEPLS.

Table 2.2 Data collection methods (targeting a total of 200 cooperative member households)

Name of study	Study dates	No. of studies	No. of respondents
FGDs	24/02/ 2020–12/ 03/2020	4 (one in each category of IPLCs: *Bawali, Mouali, Jele,* and crab collectors)	6 in each FGD
Survey	25/04/ 2021–05/ 05/2021	1	135 households (30 *Bawali*, 40 *Mouali*, 35 *Jele*, and 30 crab collectors)
PPGIS	25/08/ 2020–29/ 08/2020	1	6 from each category

Data was collected through participatory observations, focus group discussions (FGDs), a survey with a semi-structured questionnaire, and a public participation geographic information system (PPGIS) activity in two cooperatives. FGDs were conducted between 24 February and 12 March 2020 before the beginning of COVID-19 in Bangladesh. The PPGIS study was performed on 25–29 August 2020. Furthermore, the survey was conducted from 25 April to 5 May 2021 (Table 2.2). Data from the Unnayan Onneshan (UO), a Dhaka-based multidisciplinary think tank that has several biodiversity restoration and conservation programmes and has been carrying out research on the Sundarbans since 2010, was also utilised. The study therefore links numerous sources of scientific knowledge to bring forth a comprehensive and scientific understanding of ecosystem health, human health, and SEPLS management in the Sundarbans.

Members of the two cooperatives pursue their livelihoods as wood collectors (*Bawali*), fishermen (*Jele*), honey and wax collectors (*Mouali*), and crab collectors. The total number of households in the two cooperatives is 200, comprised of approximately 1500 household members. For FGDs and the survey, the households were categorised as wood collectors (*Bawali*), fishermen (*Jele*), crab collectors, and honey and wax collectors (*Mouali*). A total of four FGDs were conducted (one in each category) with six respondents in each of the FGDs. The number of households surveyed was 135 for the two cooperatives, with each participating household made up of TRUs registered as cooperative members who actively participate in SEPLS management (Table 2.2).

An area of 40 km in length and 30 km in width was selected for the PPGIS study. The region is located between 22°28′30″N and 22°1′0″N and 89°13′30″E to 89°30′0″E. This region is the part of the Khulna Range—one of the four administrative areas of the Sundarbans. Results from FGDs, the survey, and PPGIS activities have been cross-checked against supporting literature.

3   SEPLS Management: Challenges, Community Response, and Health Outcomes

The two cooperatives have been sustainably utilising and conserving the resources of the SEPLS, maintaining the well-being of both the Sundarbans and themselves. However, they have faced a plethora of challenges in managing the Sundarbans' SEPLS. In response, IPLCs have taken various actions to adapt to and mitigate these challenges (Fig. 2.1). These actions have led to (1) increased regenerative capacity of the Sundarbans and well-being of ecosystems and (2) increased income and standard of living, low-impact lifestyles, and sustainable production and consumption by IPLCs, which has contributed to positive health outcomes. On the one hand, the SEPLS remains healthy providing numerous services to the IPLCs, and on the other, IPLCs find natural solutions to the problems faced, i.e. disease and livelihood insecurity, and enjoy increased income and therefore increased expenditure on health (Fig. 2.2).

The challenges faced in SEPLS management include (a) siltation in the canals and rivers of the Sundarbans due to low flow of upstream water; (b) tidal management, water engineering, and embankments in upstream transborder waterbodies; (c) spread of invasive species; (d) climate change and salinity intrusion; (e) industrialisation and development projects near (or around) the SEPLS; (f) extracting of resources using harmful and unsustainable techniques (e.g. setting fire, poisoning water); (g) increasing habitation and illegal encroachment; (h) land shortage, land reclamation, and shrimp cultivation; (i) rent-seeking tendencies and extralegal management; (j) marginalisation of local and indigenous people and existence of poverty; (k) biodiversity degradation, frequent natural disasters, resource vulnerability, and livelihood insecurity; and (l) COVID-19.

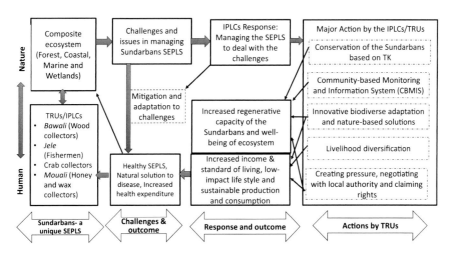

Fig. 2.2 SEPLS management: challenges, community responses, and health outcomes (source: prepared by authors)

IPLCs have adopted various innovative and participatory local approaches and actions to manage the SEPLS sustainably for ecosystem health, animal health, human health, and livelihood security. These are (a) mobilising themselves for claiming rights and protecting the SEPLS; (b) securing land through struggle; (c) negotiation with local government; (d) conservation practices based on TK; (e) community-based monitoring and information system (CBMIS); (f) community plantation; (g) homestead plantation; (h) innovative biodiverse adaptation and nature-based production, i.e. sustainable aquaculture (fish and crab culture) and sustainable forest product culture (*golpata* culture, honey and wax culture); and (i) working with local government and the forest department as a watchdog to stop illegal hunting and harvesting, cutting of trees, and usage of poison and harmful nets to catch fish. For example, the Munda Indigenous Forest People Multipurpose Development Cooperative, one of the two cooperatives, has been able to regain and restore some of its lost land amounting to 42 *bighas* (29 acres), from the powerful encroachers. The reclaimed land is located at 22°18′0″N and 89°18′10″ E (Fig. 2.3). On the other hand, the other cooperative targeted in this study, the Koyra Forest People Multipurpose Development Cooperative, has conducted community afforestation on 42.10% of 494 hectares of land along the embankment of the *Shakhbaria* River (*Rai* River) in Koyra Upazila. The area of this afforestation along the embankment is 206 hectares (Fig. 2.3). These activities enhance ecosystem health, animal health, and human health, forming a virtuous cycle. The healthy ecosystem provides more services to human beings, and thus helps promote and maintain human health.

The survey and FGDs revealed that before the onslaught of the COVID-19 pandemic, the IPLCs in these two cooperatives had been able to lead somewhat decent lives by their standards through utilising the resources of the SEPLS. They had been achieving positive outcomes from all of their actions in SEPLS management. However, the pandemic posed severe stresses on their lives and livelihoods, resulting in poor health. Cooperative members suffered a 26.16% income loss due to COVID-19 (Table 2.4). They argued that this loss of income and resulting distress in livelihoods and health were triggered by closure of economic activities and restriction of movement when the government imposed a nationwide lockdown from 23 March to 30 May 2020 due to the first wave of the pandemic. The second wave—which was supposedly deadlier—started in March 2021. All respondents argued that they were not able to go to markets to sell the resources collected from the Sundarbans. They also faced falling prices of the resources due to the pandemic. As a result, they incurred income loss. Nevertheless, they reported that during the lockdown they consumed mostly resources collected from the forest in order to cope with their livelihood hardships. The outcomes from SEPLS management have played a major role in saving lives and securing the health and livelihoods of the IPLCs in the two cooperatives. However, the COVID-19 pandemic has created new poverty, polarisation, and inequality in the society, though it has had no direct impact on the Sundarbans. The indirect impacts, however, have been significant (e.g. more harvesting to make up for loss of income).

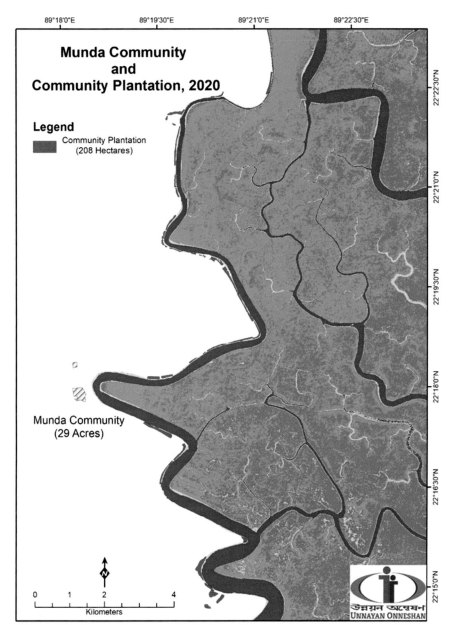

Fig. 2.3 Securing of land by Munda Cooperative and community plantation by Koyra Cooperative (source: Unnayan Onneshan, 2020a)

4 State of the Forest and Impacts of COVID-19 on TRUs

This globally acclaimed heritage site, including its sanctuaries and ecologically critical areas, which also acts as a natural wall to climatic variabilities (e.g. cyclones), is now ecologically vulnerable due to overexploitation of resources and ineffective institutions. The ill-being of the forest also negatively affects the lives and livelihoods of its people, who have suffered further hardships due to the impacts of COVID-19.

4.1 State of the Forest

Some studies have argued that even though the forest's boundary is almost unaffected, the quality of the woods is deteriorating (Hussain & Karim, 1994; Siddiqi, 2001; Iftekhar & Islam, 2004). The decadal changes in forest coverage in the part of the forest in the Khulna administrative range, drawn by PPGIS, indicate that the amount of trees is declining drastically (Fig. 2.4). As a result, the amount of fallow land is increasing. In two decades, the total area of dense forests in the case study site has halved. The coverage and density of the Sundarbans are declining.

The dark green parts in Fig. 2.4 correspond to areas of dense forest, light green parts correspond to areas of moderately dense forests, and areas where the forest is very thin are shown in white, indicating that these areas have become empty fallow lands. The white areas of fallow land have doubled in the last two decades from 4546 hectares in 2000 to 5678 hectares in 2010, and 10,501 hectares in 2020. In contrast,

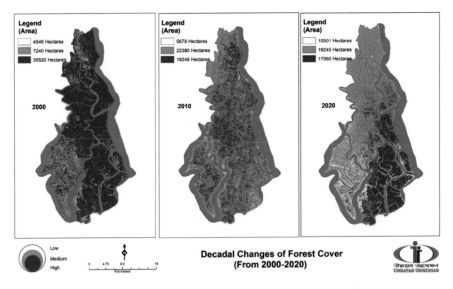

Fig. 2.4 Decadal changes of forest coverage (source: Unnayan Onneshan, 2020a)

the region of dense forests has declined sharply from 35,520 hectares in 2000 to 17,560 hectares in 2020. Similarly, there has also been a marked change in the area of moderately dense forest, increasing from 7240 hectares in 2000 to 22,380 hectares in 2010 (Fig. 2.4).

The main reasons behind the forest coverage and biodiversity loss are man-made pressures, climate change, and natural disasters (Titumir, 2021). First, illegal encroachment of forest land by powerful groups is increasing. There has been a gradual increase of human settlements around the forest. Though cutting of trees is banned, illegal tree felling is accelerating deforestation. Second, many development projects and commercial activities are being carried out, even around the ecologically critical area, therefore causing harm to the ecosystem (Titumir et al., 2020). Third, biodiversity and forest resources are being degraded as a result of over-extraction. Harmful methods of resource collection are one of the main culprits behind this loss. For example, people often use poison and harmful nets for catching fish (Titumir et al., 2019). Fourth, the frequent occurrence of catastrophic natural disasters is damaging the ecosystem and biodiversity. Fifth, climate change is negatively affecting many organic as well as inorganic components (e.g. salinity, rainfall, soil pH, mineral ingredients) of the forest (Titumir et al., 2022). Finally, existing forest law and management approaches do not recognise the traditional rights and traditional knowledge of the IPLCs. For example, TRUs need to collect a clearance certificate from the forest department to go to the forest for resource collection amounting to a certain amount of money. This system has often been accused of irregularities and corruption. These irregularities have forced TRUs to collect more resources than required to survive in order to meet extra costs (Titumir, 2021).

4.2 Impact of COVID-19 and Supercyclone Amphan on Traditional Resource Users (TRUs)

Members of the two cooperatives who depend on the forest for their livelihoods as traditional resource users have historically faced multiple pressures due to clientelist systems in forest use and management. In this context of existing hardships, the COVID-19 pandemic, nationwide lockdown, and the supercyclone *Amphan* wielded catastrophic impacts on the lives and livelihoods of many forest people. Survey results revealed that the number of households in two cooperatives suffering income loss was 103, or 76.3% of total households surveyed. The average monthly income loss was 48.98 USD (Table 2.3).

Khalil Dhali, a 53-year-old TRU and also the president of Koyra Forest People Cooperative, has been collecting resources from the forest for years. The supercyclone Amphan devastated the livelihoods of millions like Dhali, leaving wreckage in the forest:

Amphan destroyed the resources of the forest. It has swept away our home, damaged the embankments, inundated farmlands with salt water making them unfit for further cultivation.

Table 2.3 Impacts on household income during the pandemic (source: author's survey)

Total no. of households	Average no. of household members	No. of households facing income loss during the pandemic	Average monthly expenditure of households	Average monthly income before COVID-19	Average monthly income during COVID-19	Average monthly income loss due to COVID-19
135	6.5	103	17,700 BDT (212.74 USD)	15,575 BDT (187.2 USD)	11,500 BDT (138.22 USD)	4075 BDT (48.98 USD)

Table 2.4 Different impacts of COVID-19 on households

Impact/item	No. of households	Percentage
Reduction in food expenditure	15	11.1
Reduction in health expenditure	4	3.0
Reduction in clothing and shelter expenditure	18	13.3
Income loss	35	26.0
Unemployed household members	35	26.0
Households supported under government's social safety net programmes	50	37.0
Households receiving immediate relief/cash assistance from the government during pandemic	72	53.3
Households facing income loss	103	76.3
Households taking loans to bear living expenses during COVID	132	98.0
Households bearing expense by using their savings	77	57.0

Water of ponds and tube wells has become saline too. We have nothing left. We cannot even collect resources as before. Many of us are starving. There is no option left for switching the occupation currently due to coronavirus. —Khalil Dhali

The nationwide lockdown caused a severe fall in demand for the resources collected by TRUs in the Sundarbans. Accordingly, they got lower prices for their resources in the market. They could not even go to the markets for several months due to the lockdown and social distancing measures, which left most of them with no income at all. Amori Begum, a female TRU who usually goes to the forest to collect resources to contribute to her family, echoes her male counterpart:

We had to sit idle during the lockdown. We were not allowed to go to the forest and to the markets. We had no income for several months. Amphan caused another catastrophic blow on our livelihoods at the time. Even after the lockdown was relaxed, we are not getting proper prices for the resources in the market. People are less willing to buy than before.— Amori Begun

The pandemic has resulted in reduced expenditure to meet the basic needs of households (Table 2.4). Reductions in food expenditure, health expenditure, and clothing and shelter expenditure were 11.35%, 3.0%, and 13.0%, respectively.

About 26% of household members are unemployed, while 98% of households have taken loans from multiple sources such as relatives, NGOs, or informal lenders during the COVID era. About 57% of households have used their savings to bear the family expenditure (Table 2.4).

4.3 Disruption in the Ecosystem

According to the TRUs, the expropriation of resources through over-harvesting and due to numerous anthropogenic pressures is creating disruptions in the ecosystem balance, resulting in massive biodiversity degradation in the forest (Fig. 2.5) (see causes in Sect. 4.1). As a consequence, there are emergences of new diseases, and the amount and quality of ecosystem services fall, leading to livelihood distress and ill-being for the TRUs in the forest. Furthermore, according to the TRUs, the policy regime is failing to secure the jobs, food, and social security of the people adequately. Hence, the existing socio-economic distress has become dire suffering for them, which also negatively affects the ecosystem posing further pressures (Fig. 2.5).

The expropriation of the forest continues, according to TRUs, when people extract resources unsustainably to get more money in the market. It can be said, therefore, that the commodification of resources—profit-making by selling in markets—is derived from the alienation of human beings from nature. Alienation from

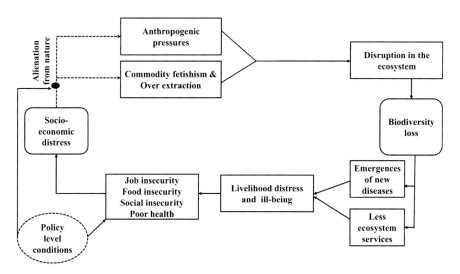

Fig. 2.5 State of the forest and its people: biodiversity loss and livelihood distress (source: prepared by authors)

nature heightens when people do not consider the true intrinsic value of nature, only monetary valuation (Titumir et al., 2019).

4.4 Policy Regime and Response to COVID Impacts

The current policy response in Bangladesh has been found inadequate to the needs of the majority of the people—informal sector workers, the poor, vulnerable, lower middle class, middle class, and other disadvantaged portions of society in face of the COVID-19. Overall, healthcare structures have been drowning in the burden of disease for months. Still now, the health sector comprises only 0.9% of GDP (Unnayan Onneshan, 2020b).

There is no provision of universal social security programmes in the country. The ongoing targeted approach—social safety net programmes—has not been able to curb the fallout from shocks, particularly for COVID-19. Existing social protection programmes are inadequate and fragmented. The selection of beneficiaries is also mired by exclusion and inclusion errors (Unnayan Onneshan, 2020b). The stimulus package announced as an immediate response to the COVID-19 pandemic has also been found ineffective for the majority of the people. The powerful and clientelist syndicate is grabbing the opportunities, while poor, vulnerable, and disadvantaged people are marginalised. The lack of an adequate and effective response from the government has heightened the suffering of IPLCs in the Sundarbans as well. Only 37% of households in the two cooperatives are covered under social safety net programmes (Table 2.4). On the other hand, only 53% of households have received immediate relief/cash assistance from the government during the pandemic (Table 2.4). Livelihood insecurity has plunged the forest people into unprecedented precarity.

5 Community Conceptualisation of the "One Health" Approach in the Sundarbans

The TRUs consider the Sundarbans to be their mind, which means they equate their lives and well-being with the life and well-being of the forest. Their thinking processes and ways of life, as well, revolve around the life and spirit of the forest. The forest contributes to the people's existence, livelihoods, breeding of their offspring, safety and security, and well-being. The IPLCs in the Sundarbans count on it. Their lives are influenced by the plethora of amenities offered by this forest, which combines numerous types of value and contributes to well-being. Well-being, as they understand it, is the health and security of both the forest and themselves, maintaining ecosystem balance. Therefore, the TRUs depend on both the biotic and

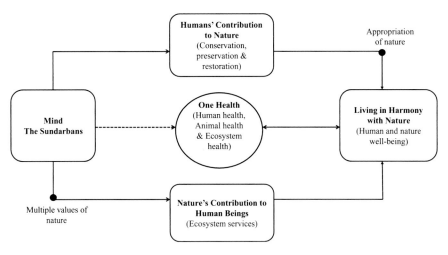

Fig. 2.6 Community conceptualisation of the "One Health" approach (source: prepared by authors)

abiotic features of the ecosystem to realise a good quality of life. Securing healthy lives for humans, animals, and the ecosystem simultaneously, as they signify, is the mainstay to realising well-being in the ecosystem. This understanding leads them to contribute to conservation, preservation, and restoration of nature, which leads to the well-being of the ecosystem (Fig. 2.6).

Therefore, the interdependent relationship between IPLCs and the Sundarbans amounts to living in harmony with nature (Fig. 2.6). In contrast, according to them, if alienation of human beings from nature prevails, commodification and thus massive extraction of resources result, leading to disruption of the ecosystem and biodiversity loss (Fig. 2.7). While IPLCs consider the Sundarbans as their life and count on the true intrinsic value of the nature, outsiders are alienated from the nature. Hence, outsiders (illegal encroachers and politically powerful business syndicates) seldom care about conserving nature.

The TRUs say that human beings often consider themselves to be "independent" or the "masters" of nature, in spite of being a part of the ecosystem. They argue that alienation from nature and treating nature as "mere matter" or as an "asset class" leads to over-extraction and destruction of nature in various ways. Therefore, human beings become separated from nature when they fail to understand the true value of it. They become unable to see themselves as part of the ecosystem and to recognise the association between humans and nature (Titumir et al., 2019). This alienation, as they suggest, results in the commodification of nature based on market-centric prices. In other words, valuation of nature becomes equal to market prices, which causes over-extraction, and thus destruction, of the natural resources (Fig. 2.7).

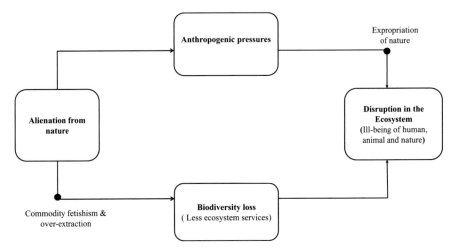

Fig. 2.7 Alienation from nature and biodiversity loss leading to ill-being (source: prepared by authors)

6 Human Contribution to Nature

In response to the continuous biodiversity loss that negatively affects the well-being and health of both the forest and its people, members of the cooperatives have adopted several innovative practices for biodiversity restoration and conservation. These practices enhance the ecosystem and animal health, which in turn promotes human health and well-being through the provision of multiple ecosystem services (Fig. 2.8).

6.1 Promotion of Customary Sustainable Practices and Traditional Knowledge

The members of the cooperatives have developed specific sustainable practices following traditional and customary knowledge. These practices are resilient and adaptive to climate change and can be promoted as innovative models for sustainable solutions for withstanding any shocks in the coastal, marine, and forest ecosystems. The communities have also developed course materials on each of the practices for training purposes and compiled an inventory on traditional knowledge (TK) with the help of UO. They apply their traditional knowledge and sustainable and customary practices for the conservation of biodiversity and nature. Further, they maintain a few specific rules and practices while harvesting resources, which are also based on traditional knowledge. They follow traditional customs and beliefs which are also consistent with resource conservation (Titumir et al., 2019). Moreover, they apply their knowledge to innovate newer techniques and methods.

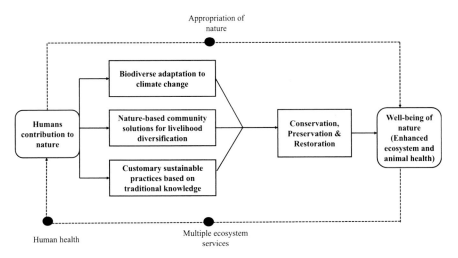

Fig. 2.8 Human contribution to nature (source: prepared by authors)

6.2 Innovations in Livelihood Options and Biodiverse Adaptation

The TRUs have diversified their livelihood choices by utilising their traditional knowledge and experiences. These practices reduce their dependence on the forest, and thus help conserve the biodiversity. Likewise, livelihood security results in good health. As alternative sources of livelihoods, they invented joint cultivation of crabs and ducks in one farmland. This practice has been found to be very profitable for the cultivators. They also developed an integrated cultivation practice for some mangrove faunal species like crabs, oyster, or fishes (e.g. *shrimps, bhetki [Latescal carifer]*) and floral species like golpata (*Nypa fruticans*), keora (*Sonneratia apetala*), and *goran* (*Ceriops decandra*) together in brackish water. This practice is known as community-based mangrove agro-aqua-silviculture (CMAAS) (Titumir et al., 2020). It serves as a substitute to commercial shrimp (CS) culture, and poses little or no negative impacts on the ecosystem (Titumir et al., 2020).

7 Modified "One Health" Approach: A Policy Perspective for the Post-2020 Biodiversity Framework

The society-wide approach to conservation and sustainable use of biodiversity through SEPLS management in the Sundarbans can ensure and enhance the "One Health" that encompasses human, animal, and ecosystem health. Promoting One Health, however, requires concerted actions. Firstly, ensuring rights-oriented social

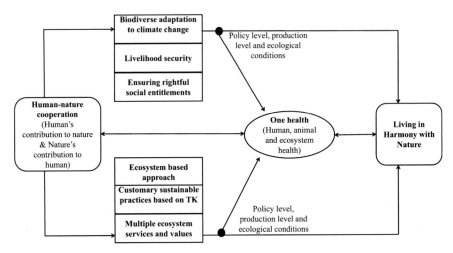

Fig. 2.9 Living in harmony with nature: modified One Health approach for post-2020 period (source: prepared by authors)

entitlement is required for everyone, such as universal access to education, healthcare, and social security programmes. Secondly, livelihood security that includes food security, job security, and social security is also essential. Thirdly, it is important to promote human-nature cooperation through green production systems such as clean and green energy, sustainable and ecosystem-based approaches to production and consumption, and biodiverse adaptation to climate change. Lastly, promotion of customary sustainable practices and traditional knowledge in SEPLS management are also required (Fig. 2.9).

These activities, however, depend on regional and global partnerships, and are also influenced at the policy level and production level and by ecological conditions. Policy-level conditions are indirect drivers of change in the ecosystem and include institutional and governance systems, power and class structures, property rights, and legal arrangements. The production level and ecological conditions are the direct drivers of change in the ecosystem. The production-level conditions include carbon emissions, pollution, over-extraction, degradation and exclusion, harvesting and fishing, reforestation, and innovative use, while the ecological factors are climate change, weather patterns, natural disasters, and other hazards. When the drivers of change (production, ecological, and political conditions) affect the ecosystem positively, the multiple services of the ecosystem contribute to human well-being. If the drivers negatively affect the ecosystem, the well-being of both humans and nature is disrupted. The appropriation of nature (conservation, restoration, sustainable use, access, and benefit sharing) ensures a harmonious relationship between humans and nature, transforming the quality of life of both. Human-nature cooperation encompasses both the human contribution to nature and nature's contribution to human beings, and contributes to maintaining a healthy ecosystem—"One Health"—leading to living in a harmony with nature (Fig. 2.9).

8 Conclusions

Exploring the interdependent relationship between humans and nature in the COVID-19 era in the Sundarbans, this case study revealed that the ecosystem in the Sundarbans is in disruption due to expropriation of nature. As a result, there has been massive degradation of biodiversity resources, which is also negatively affecting the lives and livelihoods of IPLCs. Forest coverage is decreasing, therefore negatively affecting the amount and quality of ecosystem services. The ongoing pandemic has further exacerbated the socio-economic distress of the forest people. The ill-being status of any feature in the ecosystem—biotic or abiotic—causes a metabolic rift in the ecosystem and therefore entraps the ecosystem health in a vicious cycle. The study also found that the policy response to curb the economic fallout from COVID-19 was inadequate. The absence of universal social security programmes in the midst of livelihood insecurity has heightened the suffering of the TRUs. The study also outlined that nature contributes to human well-being in multiple ways, and likewise, humans also contribute significantly to nature's well-being. Analysing the community understanding on the One Health approach demonstrated that humans and nature are dependent on each other and form a human-nature sociality in the ecosystem where they coexist. Therefore, the appropriation of nature (conservation, restoration, sustainable use, access, and benefit sharing) instead of expropriation (anthropogenic pressures) serves as a yardstick to ensure a virtuous cycle (good quality of life) in the ecosystem and the harmonious relationship between humans and nature. In this regard, a modified One Health framework for the post-2020 period was presented, which can promote a human transition to living in harmony with nature. The framework calls for ensuring rights-oriented universal social entitlements, provision of livelihood security, and promotion of human-nature cooperation through green production, ecosystem-based approaches, and biodiverse adaptation to climate change, underwritten by customary sustainable practices and traditional knowledge in SEPLS management.

References

Behera, M. D., & Haider, M. S. (2012). *Situation analysis on biodiversity conservation*. In Ecosystem for life: A Bangladesh-India initiative, IUCN, International Union for Conservation of Nature.

Díaz, S., Fargione, J., Chapin, F. S., III, & Tilman, D. (2006). Biodiversity loss threatens human well-being. *PLoS Biol, 4*(8), e277. https://doi.org/10.1371/journal.pbio.0040277

Google. (2021). Google maps [project area, Sundarbans, Bangladesh] 2021. Retrieved from https://www.google.com/maps/d/u/0/viewer?mid=1tvbwOf1OesMeRKMgqRqnL4caPP1OPs4j&ll=22.170605272107274%2C89.59105217108277&z=10.

Hussain, Z., & Karim, A. (1994). Introduction. In Z. Hussain, & G. Acharya (eds.) *Mangroves of the Sundarbans*, vol. 2: Bangladesh. IUCN, Bangkok, TH.

Iftekhar, M. S., & Islam, M. R. (2004). Degeneration of Bangladesh's Sundarbans mangroves: A management issue. *The International Forestry Review, 6*(2), 123–135.

Islam, M. S. N., & Gnauck, A. (2009). Threats to the Sundarbans mangrove wetland ecosystems from transboundary water allocation in the Ganges basin: A preliminary problem analysis. *International Journal of Ecological Economics & Statistics, 13*(9), 64–78.

Layke, C., Mapendembe, A., Brown, C., Walpole, M., & Winn, J. (2012). Indicators from the global and sub-global millennium ecosystem assessments: An analysis and next steps. *Ecological Indicators, 17*, 77–87. https://doi.org/10.1016/j.ecolind.2011.04.025

Millennium Ecosystem Assessment. (2005). *Ecosystems and human well-being: A framework for assessment*. Island Press.

New Age. (2020). Amphan causes Tk 1100 crore initial loss. Dhaka, viewed 26 July 2021. Retrieved from https://www.newagebd.net/article/106911/amphan-causes-tk-1100-crore-initial-loss.

Siddiqi, N. A. (2001). *Mangrove forestry in Bangladesh*. Institute of Forestry and Environmental Sciences, University of Chittagong.

Titumir, R. A. M., Paran, M. S., Pasha, M. W., & Meem, M. H. (2022). Climate change and its impact: Sundarbans as a natural wall. In R. A. M. Titumir (Ed.), *Sundarbans and its ecosystem services: Traditional knowledge, customary sustainable use and community based innovation*. Palgrave Macmillan. (forthcoming).

Titumir, R. A. M. (2021). Sundarbans under threat. In *Prothom Alo*, viewed 24 August 2021. Retrieved from https://en.prothomalo.com/environment/sundarbans-under-threat.

Titumir, R. A. M., Afrin, T., & Islam, M. S. (2020). Traditional knowledge, institutions and human sociality in sustainable use and conservation of biodiversity of the Sundarbans of Bangladesh. In O. Saito, S. M. Subramanian, H. Hashimoto, & K. Takeuchi (Eds.), *Managing socio-ecological production landscapes and seascapes for sustainable communities in Asia: Mapping and navigating stakeholders, policy and action* (pp. 67–92). Science for Sustainable Societies. https://doi.org/10.1007/978-981-15-1133-2_5

Titumir, R. A. M., Paran, M. S., & Pasha, M. W. (2019). The Sundarbans is our mind: An exploration into multiple values of nature in conversation with traditional resource users. In UNU-IAS & IGES (Ed.), *Understanding the multiple values associated with sustainable use in socio-ecological production landscapes and seascapes (SEPLS)* (Satoyama initiative thematic review) (Vol. 5, pp. 97–117). United Nations University Institute for the Advanced Study of Sustainability.

Titumir, R. A. M., & Afrin, T. (2017). The complementarity of human and nature well-being: A case illustrated by traditional forest resource users of the Sundarbans in Bangladesh. In UNU-IAS & IGES (Ed.), *Sustainable livelihoods in socio-ecological production landscapes and seascapes* (Satoyama Initiative Thematic Review) (Vol. 3, pp. 34–45). United Nations University Institute for the Advanced Study of Sustainability.

Unnayan Onneshan. (2020a). *Biodiversity, climate change and traditional resource users in the Sundarbans: An exploration through public participation geographic information system (PPGIS)*. Unnayan Onneshan.

Unnayan Onneshan. (2020b). *Whither Bending for life and livelihood: A rapid assessment of national budget 2020–21*, Bangladesh Economic Update, Unnayan Onneshan, vol. 11, no. 9.

Unnayan Onneshan. (2020c). Promoting diverse cultural values of biodiversity ecosystem services. In Presentation to FPP virtual partners' meeting, Unnayan Onneshan, Dhaka, Bangladesh, 11 November 2020.

van Oudenhoven, A. P. E., Petz, K., Alkemade, R., Hein, L., & de Groot, R. S. (2012). Framework for systematic indicator selection to assess effects of land management on ecosystem services. *Ecological Indicators, 21*, 110–122. https://doi.org/10.1016/j.ecolind.2012.01.012

Wikimedia Commons Contributor Nirvik12. (2015). Map of Sundarbans. Wikimedia Commons, the free media repository, Public Domain, viewed 16 December 2021. Retrieved from https://commons.wikimedia.org/wiki/File:%E0%A6%B8%E0%A7%81%E0%A6%A8%E0%A7%8D%E0%A6%A6%E0%A6%B0%E0%A6%AC%E0%A6%A8%E0%A7%87%E0%A6%B0_%E0%A6%AE%E0%A6%BE%E0%A6%A8%E0%A6%9A%E0%A6%BF%E0%A6%A4%E0%A7%8D%E0%A6%B0.svg.

The opinions expressed in this chapter are those of the author(s) and do not necessarily reflect the views of UNU-IAS, its Board of Directors, or the countries they represent.

Open Access This chapter is licenced under the terms of the Creative Commons Attribution 3.0 IGO Licence (http://creativecommons.org/licenses/by/3.0/igo/), which permits use, sharing, adaptation, distribution and reproduction in any medium or format, as long as you give appropriate credit to UNU-IAS, provide a link to the Creative Commons licence and indicate if changes were made.

The use of the UNU-IAS name and logo, shall be subject to a separate written licence agreement between UNU-IAS and the user and is not authorised as part of this CC BY 3.0 IGO licence. Note that the link provided above includes additional terms and conditions of the licence.

The images or other third party material in this chapter are included in the chapter's Creative Commons licence, unless indicated otherwise in a credit line to the material. If material is not included in the chapter's Creative Commons licence and your intended use is not permitted by statutory regulation or exceeds the permitted use, you will need to obtain permission directly from the copyright holder.

Chapter 3
Linking Biocultural Memory Conservation and Human Well-Being in Indigenous Socio-Ecological Production Landscapes in the Colombian Pacific Region

Andrés Quintero-Angel, Andrés López-Rosada, Mauricio Quintero-Angel, David Quintero-Angel, Diana Mendoza-Salazar, Sara Catalina Rodríguez-Díaz, and Sebastian Orjuela-Salazar

Abstract The Colombian Pacific region is one of the most biodiverse areas in the world; however, it is severely threatened by anthropogenic pressures. In addition, armed conflict and poverty are compounding factors causing the loss of biodiversity and cultural identity. In response to this situation, the *Wounaan-Nonam* original people of Puerto Pizario and Santa Rosa de Guayacán declared five Indigenous Protected Areas (IPA) in 2008. We conducted a study to highlight the link between the conservation of biocultural memory and contributions to human well-being, particularly to human health, in indigenous socio-ecological production landscapes and seascapes (SEPLS). Since 2013, the research-action-participation methodology has been applied to recover ecological traditional knowledge on how ancestors managed nature and elements associated with their cosmovision. Following the TNC conservation of areas methodology, eight biological and cultural conservation values were identified for the IPAs and 5-year management plans for conservation were formulated. As a result of this process, we created a tool that involves traditional knowledge to administer the total 1850 hectares covered by the five

A. Quintero-Angel (✉)
Corporación Ambiental y Forestal del Pacifico—CORFOPAL, Cali, Valle del Cauca, Colombia

Asociación de Cabildos Indígenas del Valle del Cauca ACIVA-RP., Buenaventura, Valle del Cauca, Colombia
e-mail: direccioncientifica@corfopal.org

A. López-Rosada
Asociación de Cabildos Indígenas del Valle del Cauca ACIVA-RP., Buenaventura, Valle del Cauca, Colombia

M. Quintero-Angel · D. Mendoza-Salazar
Universidad del Valle, sede Palmira, Palmira, Valle del Cauca, Colombia

D. Quintero-Angel · S. C. Rodríguez-Díaz · S. Orjuela-Salazar
Corporación Ambiental y Forestal del Pacifico—CORFOPAL, Cali, Valle del Cauca, Colombia

© The Author(s) 2022
M. Nishi et al. (eds.), *Biodiversity-Health-Sustainability Nexus in Socio-Ecological Production Landscapes and Seascapes (SEPLS)*, Satoyama Initiative Thematic Review, https://doi.org/10.1007/978-981-16-9893-4_3

IPAs. We also found that the main challenges faced by indigenous communities in the management of IPAs as an integral part of the indigenous SEPLS are associated with weak organisational and governance processes. Additionally, we identified the main opportunities ecosystem services offer in the IPAs, which enhance the quality of life and health of the original peoples and ecosystems at a regional level. Finally, the making of handicrafts is identified as an opportunity in these SELPS, as it represents an alternative for generating income through sustainable productive chains in biotrade strategies.

Keywords Biocultural memory · Human health · Indigenous Protected Areas · SEPLS · Heritage conservation

1   Introduction

When considering that humans are just one species among the vast natural diversity of the planet, Toledo and Barrera (2008) state that the success of human survival lies in ecological factors such as population size, evolutionary processes, brain capacity, and, with it, memory. These elements have made it possible to achieve different cultural processes of "biological, genetic, linguistic, cognitive, agricultural, and landscape" diversification (Toledo & Barrera, 2008, p. 25), based on a necessary recognition and appropriation of nature. Thanks to human capability and as a socio-historical process, communities around the world have taken advantage of the particularities of varying landscapes in their local environments according to their material and spiritual needs (Lindholm & Ekblom, 2019; Ekblom et al., 2019; Toledo et al., 2019).

In this context, Toledo and Barrera (2008) point out that indigenous peoples[1] maintain and/or possess at least 80% of the planet's cultural diversity. They also posit that the presence, permanence, and resistance of these communities in ancestral territories have been fundamental for the conservation of ecosystems. These indigenous territories can be considered socio-ecological production landscapes and seascapes (SEPLS) as they allow for a socio-ecological balance in the productive use of landscapes based on biocultural memory.[2] These types of SEPLS are present

---

[1] Indigenous people are composed of communities which "… having a historical continuity with pre-invasion and pre-colonial societies that developed on their territories, consider themselves distinct from other sectors of the societies now prevailing on those territories, or parts of them. They form at present non-dominant sectors of society and are determined to preserve, develop and transmit to future generations their ancestral territories, and their ethnic identity, as the basis of their continued existence as peoples, in accordance with their own cultural patterns, social institutions and legal system" (United Nations, 2004: 2). However, by their own identification, they are now referred to in Latin America as "original people" (*pueblos originarios*).

[2] "The concept of biocultural memory has been used as a synonym of accumulated ecological knowledge or social-ecological memory, referring to ecological knowledge and practices that are regenerated, retained, and revived through collective memory by communities of users" (Garavito-Bermúdez, 2020, p. 4–5).

in the Pacific region of Western Colombia, where numerous ethnic groups live (Departamento Administrativo Nacional de Estadística (DANE), 2019) in one of the most biodiverse areas in the world (Losos & Leigh, 2004; Plotkin et al., 2000; Rangel et al., 2004).

As a product of biocultural memory, the original people of the Colombian Pacific region, such as the *Wounaan-Nonam*, maintain a holistic vision of their territorial heritage, whereby any environmental problem or conflict is considered as an imbalance and illness of Mother Earth (or *Pachamama*). In this regard, the concept of health extends to animal, plant, and human health. Hence, various sustainable practices associated with local ancestral knowledge for healthcare and well-being endure, for example, the implementation of harmonisation rites for the use of natural resources to maintain the ecosystem balance, spiritual cleansing with sacred plants, and healing rituals. For these original people, well-being is synonymous with health; for this reason, well-being (or *Sumak Kawsay* as it is called in other Andean indigenous communities) is sought in all aspects of life, and maintaining a communal state of well-being means enjoying harmony and health.

Despite its great biological and cultural diversity, the Colombian Pacific region is threatened by a trend towards agro-industrial modernisation (Quintero-Angel et al., 2020) and pressures from illegal armed groups and drug trafficking, which have put the survival of different ethnic groups at risk (Vélez et al., 2020; Defensoría del Pueblo, 2018; Ministerio de Justicia & UNDOC, 2013). Particularly, the *Wounaan-Nonam* of Puerto Pizario's reservations, located on the banks of the San Juan River, and of Santa Rosa de Guayacán on the Calima River Basin, face these tensions and are trying to conserve the biocultural memory of their territories. One of the strategies employed for the conservation of *Wounaan-Nonam* territories, which are traditionally managed as SEPLS, has been the declaration of Indigenous Protected Areas (IPAs). In this context, the present case study aims to demonstrate the link between conservation of biocultural memory and its contributions to human well-being and health, taking as an example the *Wounaan-Nonam* SEPLS. In this sense, we claim that this case study is a joint construction with the indigenous authorities of these reservations. All information presented has been reviewed and endorsed by them with their prior, informed consent, and more detailed quantitative data on these SEPLS, as well as their properties and uses, remains the confidential information of the indigenous communities.

1.1 Current Situation of the Original People of Colombia and the Wounaan-Nonam *Indigenous Communities*

Since the beginning of the twenty-first century, Colombia has recognised the historical exclusion and vulnerability in which original people have lived, to the extent that today they are at risk of physical and cultural extermination (Decision

T-025 of 2004). In Colombia there are 115 original peoples with an approximate population of 1,905,617, grouped in 710 reservations[3] in the national territory, where 65 native languages are spoken (Departamento Administrativo Nacional de Estadística (DANE), 2019). Original people in Colombia live in extreme conditions of human right violations due to armed conflict, which has generated forced displacement migrations, along with the deterioration of natural resources and severe social exclusion (Mamo, 2020); these reasons explain why diverse ethnic groups are currently dispersed in areas far from their original settings.

These difficult living conditions of original people in Colombia, and around the world, are directly related to environmental conflicts with the state, individuals, multinational corporations, and megaprojects that intervene in their territories by means of using their soil for illicit crops and subsoil for mining activities, dam construction, or agro-industrial activities (Semana Sostenible, 2019). In the Colombian context, illicit crops and drug trafficking are additional factors that come into play. Consequently, these communities have called for the joint design of strategies originated within indigenous communities (Ministerio de Cultura, 2009), to document and execute their Life Plans[4] and Safeguard Plan[5] and to generate studies and research to recover and strengthen the ancestral traditional ecological knowledge (TEK) of each indigenous community.

The *Wounaan-Nonam* community is composed of approximately 14,825 people (Departamento Administrativo Nacional de Estadística (DANE), 2019). Its original territory was located next to the San Juan River, between the Chocó and Valle del Cauca departments, a forested piedmont they share with the *Embera* and *Kuna* ethnic groups. Unfortunately, the *Wounaan-Nonam* people are one of the 34 original peoples at risk of disappearing due to forced displacement caused by armed conflict in Colombia. A decision of the Constitutional Court, Auto 004 of 2009, recognised this reason for the dispersal across the country far from their ancestral territories (Organización Nacional Indígena de Colombia, 2018). The fundamental individual and collective rights of the people have been taken away, affecting their cultural autonomy and identity, and putting them at risk of physical and cultural extinction.

---

[3] "Reservations are legal institutions of sociopolitical special character, composed of one or more indigenous communities which have a collective property title with the same guarantees of private property, they have their territory, and they manage it and their internal life by an autonomous organisation covered by the indigenous status and their own normative system" (Decree 2164 of 1995, art. 21).

[4] Planning instrument which involves a document with diagnostic information to deal with the necessities and challenges of each indigenous community from their own cosmovision, facilitating dialogue with local and national governments to address inclusion and governance processes.

[5] The Safeguard Plan is defined as a social and administrative agreement that establishes directions, recommendations, and actions to guarantee and safeguard the intangible cultural heritage and national patrimony (Ministerio de Cultura, 2009).

1.2   Study Area: Indigenous Reservations of Puerto Pizario and Santa Rosa de Guayacán

The indigenous reservations of Puerto Pizario and Santa Rosa de Guayacán are located in the Pacific region of Colombia, in the basins of two large rivers, the San Juan (Valle del Cauca and Choco departments) and Calima (Valle del Cauca department), respectively (Fig. 3.1, Table 3.1).

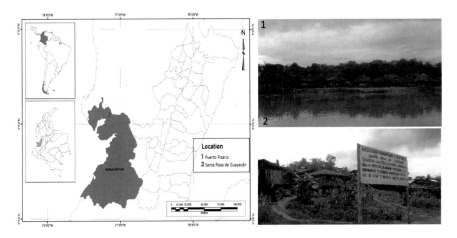

Fig. 3.1 Location of (1) Puerto Pizario and (2) Santa Rosa de Guayacán indigenous reservations. Photo credits and map by López-Rosada

Table 3.1 Basic information on the study area

Study site	Puerto pizario indigenous reservation	Santa Rosa de Guayacán indigenous reservation
Country	Colombia	
Province	Chocó/Valle del Cauca	Valle del Cauca
Municipality	Litoral del San Juan/Special District of Buenaventura	Special District of Buenaventura
Dominant ethnicity, if appropriate	Indigenous and Afro-Colombian communities	
Case study/project area size (hectare)	2920	236
Number of direct beneficiaries	789	155
Dominant ethnicity in the project area	*Wounaan-Nonam* indigenous ethnicity	
Geographic coordinates (latitude, longitude)	4°13′31.16″ N 77°15′18.78″ W	4°6′55.79″ N 77°8′36.50″ W

Indigenous Reservation of Puerto Pizario

The settlement was founded in 1975 and declared an Indigenous Reservation with Resolution No. 013 of 1983 from the Instituto Colombiano de Reforma Agraria (INCORA). It is located on both banks of the San Juan River and covers 2920 ha, of which the largest part is in the Chocó department. It is a terrain dominated by hills and dense humid rainforest, and home to 789 *Wounaan-Nonam* inhabitants, who have declared three IPAs since 2008 (Fig. 3.2a, Table 3.2).

Humanitarian and Biodiverse Indigenous Reservation of Santa Rosa de Guayacán

The settlement was established in 1979. It became a reservation through the INCORA Resolution No. 054 of 1989 a decade later. It extends for 236 ha, is inhabited by 155 *Wounaan-Nonam*, and contains two IPAs, declared in 2008. It acquired its humanitarian and biodiverse reservation status when the territory was returned to the original people on 14 August 2010, prior to which they had become victims of forced displacement. This should be highlighted as a form of resistance to the threats of armed conflict and the struggle to maintain their lifestyle and ancestral customs (Fig. 3.2B, Table 3.2).

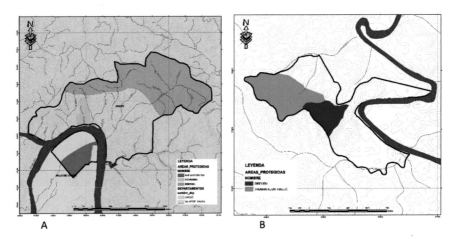

Fig. 3.2 (a) Location of the Indigenous Protected Areas of Puerto Pizario and (b) Santa Rosa de Guayacán reservations. Maps by López-Rosada based on Quintero-Ángel et al., 2015

Table 3.2 Description of the Indigenous Protected Areas of Puerto Pizario and Santa Rosa de Guayacán reservations

IPA name	Indigenous reservation	Area (ha)	Location	Description
Nueva Floresta	Puerto Pizario	212	Valle del Cauca	Flat zone, occasionally flooded by the San Juan River, with some human interventions such as selective wood extraction. Some big native trees remain, and some species of fauna are used by communities as protein sources
Pizabarra	Puerto Pizario	103	Chocó	Covers the Pizabarra watershed, from its source to the San Juan River outlet. There is slight intervention in the lower part by logging and extraction of handicraft materials. The conservation status of the rest of the watershed is considered good
Beermia Native language meaning: "white-collared peccary spirit (*Pecari tajacu*)"	Puerto Pizario	1446	Chocó	Equivalent to 32% of the reservation, it covers the higher part of the area and is the source of the Cuellar, Medio, Llano, and Dupurma streams. It is in very good conservation status. It has a high quantity of flora and fauna highlighted by the presence of white-collared peccaries (*Pecari tajacu*), jaguars (*Panthera onca*), and deer (*Mazama americana*), as well as timber species that are less common in other areas
Dibeeudu Native language meaning: "place of many chonta palms"	Santa Rosa de Guayacán	23.4	Valle del Cauca	Corresponds to the Micurero watershed. It is characterised by steep slopes and forests in good conservation status. It borders family productive areas in the lower part
Thumaan Kuun Khirjug Native language meaning: "collective thinking"	Santa Rosa de Guayacán	57	Valle del Cauca	Located in the upper part of the Cienaga and Cienaguita watersheds. The area is covered by well-preserved forests, except for the area adjacent to the road that goes to the Bahía Málaga Naval Base, where foreigners have been logging and poaching

2 Methodology

The methodology was developed in five phases (Fig. 3.3) throughout 2013 by implementing specific actions in Spanish and the native language, including (1) key informant interviews (Geilfus, 2002); (2) participant observation (Kawulich, 2005); (3) focus groups of men, women, youth, and children (Geilfus, 2002); (4) social cartography (Geilfus, 2002); and (5) exchange of knowledge (PRATEC, 2012) (Fig. 3.4). These actions were developed during several meetings and workshops with the communities, where they shared their ancestral knowledge, recovering and documenting wisdom on how their ancestors managed nature, agricultural practices, and innovations to improve food security, heirloom practices of traditional medicine, health status of the community and the environment, handicraft fabrication, and elements associated with indigenous spirituality[6] and cosmovision.

Finally, through the research-action-participation methodology, exchange of knowledge (PRATEC, 2012) and analytical triangulation (Rodríguez-Sabiote et al., 2006) were carried out to establish the planning of Indigenous Protected Areas in each reservation. In order to do this, the planning for conservation of areas methodology was also followed (Granizo et al., 2006), and biological and cultural objects of conservation were identified and prioritised. A viability analysis was conducted, and management plans for conservation were formulated, which include monitoring and follow-up indicators for 5 years in each of the areas with each indigenous community.

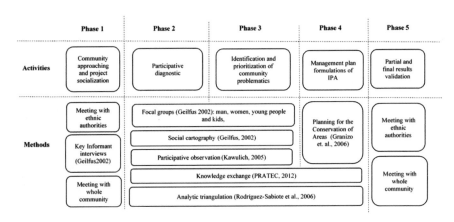

Fig. 3.3 Visual presentation of methods and activities (prepared by authors)

---

[6]Indigenous spiritual practices are the belief in something beyond an individual connection to others and to the world and their close relationship with a higher power. They are part of the resistance to the effects of colonialism (Consejo Regional Indígena del Cauca (CRIC), n.d.).

Fig. 3.4 (**a**) A team member working with a focus group of kids on the fauna present in Santa Rosa de Guayacán; (**b**) original people from Puerto Pizario explaining the results of social cartography; (**c**) knowledge exchange exercise in Santa Rosa de Guayacán; (**d**) socialisation of knowledge exchange exercise in Puerto Pizario (photo credits: A. Quintero-Angel 2014)

3   Results and Discussion

3.1   Cosmovision and Relationship with Nature of **Wounaan-Nonam** *Indigenous Communities*

Inside of the *Wounaan-Nonam* cosmovision, natural elements such as water, movement, and earth are the principles of the Origin Law[7] that give life to "*Maach Aai*" (elderly father-God) and to all the spiritual elements, gods, male and female principles,[8] sacred sites, material goods, and natural richness. Initiation and life rituals revolve around traditional elements such as body paintings with organic dyes,

---

[7] Origin Law are the laws for cultural identity of indigenous peoples, ancestral laws, and fundamentals (Consejo Superior de la Judicatura (ONIC), 2011, pp. 256–259).

[8] Inside this culture women are highlighted and have an important role as they are in charge of oral passing on of custom, stories, fabrics, spirituality, and socialisation of children in the communities.

Fig. 3.5 *Wounaan-Nonam* handicraft sample made from Mocora palm (*Astrocaryum standleyanum*) and natural dyes (photo credits: López-Rosada, 2016)

fabrics made from natural materials, dances, and songs. These express the cosmovision and religion of this ethnic group and its close relation with nature (Ministerio de Cultura, 2010; Ministerio del Interior, 2012).

In *Wounaan-Nonam* cosmogony, the whole ecosystem relates to the spiritual component through "*Chimias*" (a spirit present in all living and nonliving things) and consequently, all the territory is related to spirituality, health, and human well-being. The ancestral link between the community and natural resources makes many species an important part of the traditional knowledge of the *Wounaan-Nonam* biocultural diversity because the diversity of medicinal plants and animals used can strengthen the spirits of human beings. Therefore, it is important to maintain the spirits (*Chimias*) of all the different species who live inside the territory. Moreover, these fauna and flora are also used in cultural practices that benefit the community (e.g. natural dye is used both in textile handicrafts and in body painting).

Wounaan-Nonam original people have maintained a harmonious relationship with nature due to their small population size and level of understanding of their natural surroundings, including the forest (e.g. meat, fruits, materials), river (e.g. fishing, transport, water), and small swidden agriculture. Although these communities live off a subsistence economy based on crops and small-scale production, unlike other ethnic groups, handicrafts with "*werregue*" Mocora palm (*Astrocaryum standleyanum*) (Fig. 3.5) and use of natural dyes such as "*achiote*" (*Bixa orellana*) have developed more at the productive level due to high commercial acceptance, which could become an alternative to generate income (Reyes-Ardila, 2019). However, the production and commercialisation of such goods must be done through sustainable production chains based on fair biotrade strategies. In addition to being a potential income source, these practices contribute to the conservation of biocultural memory, given that stories of mythical characters, animals, and situations of daily life are captured in them.

The use of a broad list of species implies an ancestral knowledge of land use that ranges from crop rotation to leaving land fallow, and land-clearing practices that leave a good number of trees standing, where the planting of banana (*Musa acuminata*), maize (*Zea mays*), manioc (*Manihot esculenta*), and sugar cane (*Saccharum officinarum*) stands out. Hunting is restricted to what is necessary, without overexploiting resources; however, population growth within the reservations and foreign intrusion threaten the permanence of some species used by the communities in the reservations. In general, the records of fauna and flora species provided by the communities are biased towards the species that are used or have a direct relationship with the communities. For example, the species recorded, such as Baudo guan (*Penelope ortoni*), yellow-throated toucan (*Ramphastos swainsonii*), collared aracari (*Pteroglossus sanguineus*), and great curassow (*Crax rubra*) for birds, and white-lipped peccary (*Tayassu pecari*), red brocket deer (*Mazama americana*), paca (*Cuniculus paca*), and white-collared peccary (*Pecari tajacu*) for mammals, correspond to those that could be used as food sources or that are common in settlements, along the riverbanks, or in crop areas. Species that are either less often seen or very small, such as amphibians and reptiles, are not given special names, which makes identification difficult.

In these original people communities, the health of inhabitants has customarily been the result of harmonious relationships established in the use of natural resources by their medical authorities to provide solutions for deficiencies of body and spirit. For example, there are diseases caused by imbalance in the spiritual world due to the absence or presence of bad or good *Chimias*, which must be treated through spiritual practices such as prayers, songs, and rituals. Likewise, physical diseases caused by the environment, such as stomach problems caused by parasites, are treated with medicinal plants. Therefore, the *Wounaan-Nonam* people contemplate environmental health under the principle of harmony of individuals and communities with Mother Earth (*Pachamama*). Certain unbalances with regard to this harmony are treated by doctors or wise persons possessing the knowledge of traditional medicines that heal both soul and body. Biodiversity and well-being interconnect with the multiple dimensions of health (e.g. physical, mental, and spiritual) through wisdom, knowledge, ancestral memory, spirituality, and relationship with Mother Earth. These elements represent the use of interculturality as a tool for the management of indigenous SEPLS and converge in indigenous thinking through the concepts of Origin Law, self-government,[9] and Greater Right[10] (Walsh, 2008). This is evidenced in the particularities of the different roles of each of the traditional spiritual authorities (Table 3.3), which transcend the established Western rationality and blur gender lines, but are fundamental for the better understanding of their world view and cosmogony.

---

[9]The United Nations Declaration on the Rights of Indigenous Peoples (United Nations, 2007) stipulates their right to autonomy or self-government in matters relating to their internal and local affairs, as well as to provide the necessary means for financing their autonomous functions (Art. 4). Therefore, the concept of autonomy also includes the relationship with the state, and how to obtain economic resources through revenue sharing.

[10]The Greater Right or *Derecho mayor* (in Spanish) is a right associated with the heritage of ancestors who lived in certain territories centuries ago. Therefore, it is important to describe this notion keeping in mind that it is essentially the right to a territory and its management by a people's own laws (Flórez-Vargas, 2016).

Table 3.3 Medical and spiritual authorities that stand out in the *Wounaan-Nonam* communities (source: prepared by author)

Traditional spiritual authorities (spiritual medics or shamans)	Description
Benkhuum	The highest medical and spiritual authority, his knowledge of their territory and resources (animals and plants) allows him to make corporal treatments, and his knowledge of the spiritual world and the handling of *Chimias* allows spiritual treatments
Pildecero or *Tonguero*	He diagnoses illnesses of the body and spirit by consuming plants such as Pildé (*Banisteriopsis caapi*) whose properties give him the power to transcend materiality and know the patient's condition
Yerbatero	Has the features of a botanist, i.e. specialises in the knowledge of plants' potentialities and the treatments that should be carried out using them
Sobandero	Has knowledge about the functioning of bones and muscles in the human body, and thus the ability to treat and rehabilitate traumas

3.2 Latent Threats to the Biocultural Memory of Wounaan-Nonam *Original People*

Wounaan-Nonam territories constitute strategic corridors for illegal armed groups as well as for planting, processing, and transportation of illicit crops. These illegal armed groups have inflicted harm on the health of the environment by imposing certain limits to the indigenous territory and preventing the free movement of people and within it, by limiting the access to certain foodstuff, and through the implementation of some foreign agricultural practices. The productive activities are restricted by the presence of dangerous anti-personnel mines in the soils for crops, hunting, and fishing areas (Defensoría del Pueblo, 2018; Ministerio de Justicia & UNDOC, 2013). Additionally, several environmental changes are associated with the El Niño Southern Oscillation (ENSO), as well as with the effects of mining, logging, and river pollution. Moreover, glyphosate spraying in the eradication programmes of illicit crops affects the soil and water and threatens their food and access to a balanced diet, as they are restricted mainly to rice and plantains. As a result, the original people have lost food autonomy[11] and they now depend on imported rice, a staple they do not produce in their territory. Consequently, in recent years food insecurity has been a critical factor for the *Wounaan-Nonam* people, with 93% of the inhabitants considering their diet as fair or poor. This is oxymoronic in the context of great biodiversity, and therefore, agrodiversity could contribute to food security and sovereignty (Burlingame et al., 2012), as well as good health.

---

[11] Food autonomy is defined by Morales (2012, p. 3) as the "right that assists each community, people or human collective, member of a nation, to autonomously control their own food process according to their traditions, uses, customs, needs and strategic perspectives, and in harmony with other human groups, the environment and future generations".

These latent threats in the territory also affect their survival as an ancient culture, their traditional medicine, customary practices, human rights guarantees (Ministerio del Interior, 2012), and biocultural memory. According to Ministerio del Interior (2012), at least 79% of inhabitants of the *Wounaan-Nonam* communities consider that their traditional medicine has been affected by threats to the lives of traditional doctors and the restrictions within the territories that prevent access to medicinal and sacred plants (Ministerio del Interior, 2012). In this sense, it is important to promote practices and incentives to support the conservation of biocultural memory in these communities, given that the loss of cultural values is linked to the degradation of natural resources (Lopez-Maldonado & Berkes, 2017; Toledo & Barrera, 2008).

Another important element to the conservation of biocultural memory is language. The majority of the *Wounaan-Nonam* are multilingual, using the native language *"Woun meu"* (belonging to the lingual family of Chocó) and Spanish. The absence of an indigenous educational system (SEIP) after primary school, generational change, and outmigration are aspects that have slowly influenced the loss of their language; in Colombia alone, indigenous language loss is higher than 50% (Departamento Administrativo Nacional de Estadística (DANE), 2019).

Despite this trend, since 2009, in the frame of the *Wounaan-Nonam* People Congress, language has been highlighted as an identity element that allows the development of resistance and community cohesion strategies and has been deemed as a priority for these people (Ministerio de Cultura, 2011). The *Wounaan-Nonam* people, like other original people in the country that preserve their native language, consider language as an essential element in terms of culture, representation, and identity, as it enables the intergenerational passing of their ancestral knowledge on environmental health as biocultural memory.[12]

Overall, armed conflict and contact with other populations have generated a cultural fragmentation process, which gravely endangers traditional management of natural resources. Additionally, lack of intergenerational communication and transmission of traditional knowledge for risk management of the SEPLS has hindered effective conservation processes in these territories. Therefore, several urgent actions are required from the Colombian Government to conserve the biocultural memory of these original people and to promote actions in accordance with their precarious conditions, cultural expectations, and survival needs. Furthermore, an improvement in equity, empowerment, and good governance of these communities and their traditional authorities (self-government and indigenous guard[13]) are fundamental in these SEPLS, given the importance of these factors

---

[12] Since the 1950s, there have been studies conducted on the *Woun meu* language that allowed the *Wounaan-Nonam* to go from oral language to a writing system. See Mejía, 2000.

[13] The Indigenous Guard is conceived as an ancestral organism and an instrument of resistance, unity, and autonomy to defend the territory and way of life for indigenous communities. It is not a police force, but a humanitarian and civil resistance mechanism. It aims to protect and disseminate ancestral culture and exercise their own rights. Its mandate is derived from their own meetings; therefore, it depends directly on indigenous authorities. Members of the Indigenous Guard rise to defend the communities from other stakeholders who may attack their people, but their only defense

for biocultural heritage conservation (Oldekop et al., 2016; Sarmiento & Viteri, 2015; Speelman et al., 2014).

3.3 Implemented Actions for the Conservation of Biocultural Memory in Indigenous Protected Areas of the Wounaan-Nonam

Despite the great threats to the biocultural memory of the *Wounaan-Nonam*, multiple cultural practices associated with ancestral knowledge (e.g. personal health, handicrafts, body painting, language, and clothes, among others) are found throughout their territory thanks to the management of the SEPLS, which has been developed in the search for harmony with nature. Therefore, the IPAs in their territory are key elements for biocultural memory conservation and a framework for improving human well-being through ecosystem-based management. They serve as reservoirs of important ecological and cultural services such as medicinal and/or spiritual plants, raw materials for handicrafts, construction materials for housing, firewood (for domestic consumption), animal species (i.e. protein sources), and fish, a central element of these communities' diet which is sometimes associated with sacred sites.[14] These sites increase protection and offer conservation opportunities (eds. Sarmiento & Hitchner, 2019; Dudley et al., 2009), as evidenced inside the Pizabarra IPA where the community cemetery is considered a sacred site as the resting place of their ancestors.

Considering the tight-knit relationship these communities share with nature due to their cosmogony, and the importance of their native language, during workshops and knowledge exchange sessions, emphasis was placed on recovering the vernacular names of fauna and flora species and the cultural elements present in both communities. In all, we were able to recover traditional knowledge on 122 species of fauna, 144 flora species, and 40 cultural elements, all associated with customary ways of life and ancestral use of these indigenous SEPLS. In order to disseminate this knowledge to the youth of the Puerto Pizario and Santa Rosa de Guayacán communities, a pedagogical tool was developed, which through a simple game allows the exchange of knowledge and contributes to the conservation of the Indigenous Protected Areas declared in these territories.[15]

---

is their baton, or "*chonta*" (maximum symbol of indigenous leadership), which gives a symbolic value to the guard (Consejo Regional Indígena del Cauca (CRIC), n.d.).

[14] Sacred sites are areas of land or water that have a special spiritual meaning for indigenous communities. They can include mountains, hills, forests, woods, rivers, lakes, islands, mangroves, streams, and caves (Oviedo & Jeanrenaud, 2007; Wild & McLeod, 2008).

[15] For detailed information about the species of fauna and flora and cultural elements recovered and used in the pedagogic tool, please see Quintero-Ángel et al. (2015).

Additionally, together with the indigenous communities, the conservation objectives of the IPAs were prioritised based on the system of protected areas in departmental planning (Corporación Autónoma Regional del Valle del Cauca (CVC), 2007). Then, we proceeded to select the tangible and intangible values of conservation objects, although due to the high biological and cultural diversity in these territories, a wide-ranging list of the possible values of conservation objects was identified (73 biological and 16 cultural). Hence, a prioritisation analysis was carried out following Granizo et al.'s (2006) approach, resulting in the selection of eight cultural and biological conservation objects for the IPAs (Table 3.4, Fig. 3.6).

Based on the selection of these values of conservation objects and following the guidelines set forth by Borrini-Feyerabend et al. (2004), a management plan for these areas was developed in a technical document with six interrelated components: (1) characterisation of the communities of Puerto Pizario and Santa Rosa de Guayacán, including their socio-economic activities; (2) participatory diagnostic of the territory (biophysical aspects); (3) administrative structure of self-government; (4) rules and safeguards of ancestral territory; (5) indicators for follow-up and monitoring; and (6) strategic planning and monitoring establishing goals to be achieved between 2014 and 2019. These goals address the following issues: conservation and restoration, ancestral use and management of biodiversity with its ecosystem services, traditional knowledge, scientific research articulated with knowledge exchanges, and empowerment. The components are articulated within a legal and conceptual framework protected by the Law of Origin, the Major Indigenous Law, and the Law 21 of 1991 which ratified for Colombia the International Labour Organisation (ILO) Convention 169 of 1989 concerning indigenous and tribal peoples in independent countries.

The establishment of IPAs administered under management plans not only is beneficial for biodiversity conservation, but can also have positive effects on various aspects of human well-being, including physical and mental health benefits associated with the relationship indigenous communities have with nature. IPAs also provide important ecosystem services such as food provisioning or medicinal resources, regulation services (e.g. pest and disease control), and aesthetic and spiritual services that favour the well-being of these communities.

4 Key Aspects of Biocultural Memory Conservation and Relationship with Health and Human Welfare

The concept of biocultural memory refers to the collective construction of knowledge associated with nature, its conservation, and intergenerational transmission, whose axis is the cosmovision and cultural representation present in each original people. In southwestern Colombia, the *Wounaan-Nonam* original people seek to maintain a balanced relationship with nature, and this is represented in spiritual and physical harmony; any imbalance is considered a disease in human beings,

Table 3.4 Conservation object values prioritised in environmental management plan

Conservation objectives	Specific objectives	Type of object	Values of conservation objects	Justification
	"Preserve and restore the natural condition of spaces that represent Colombian ecosystems or characteristic combinations of them"		Ecosystems (warm pluvial forest in fluvial-marine plain (BOCPLRY))	This ecosystem has zero representation in protected areas in the Valle del Cauca department. Its extension is 10,838.5 ha, which represents 0.5% of the total area of Valle del Cauca, making it an important value of conservation object (Corporación Autónoma Regional del Valle del Cauca (CVC) and Fundación Agua Viva (FUNAGUA), 2010)
			Fine timber: Carrá (*Huberodendrom patinoi*), Chachajo (*Aniba perutilis*), Chanul (*Humiriastrum procerum*)	Selected due to the high pressure these species are under in IPAs and in general in the Colombian Pacific due to their high commercial value (eds. Cárdenas & Salinas, 2007)
"Ensure the continuity of ecological and evolutionary processes to maintain biological diversity"	"Preserve populations and habitats necessary for the survival of wild species or sets of species that present particular conditions of special interest for the conservation of biodiversity, with emphasis on those of restricted distribution"	Biological	Large mammals: White-lipped peccary (*Tayassu pecari*), red brocket deer (*Mazama americana*), paca (*Cuniculus paca*), white-collared peccary (*Pecari tajacu*), and jaguar (*Panthera onca*)	Group of a large number of species of high ecological importance for the ecosystems and for the community, due to the protein contribution that these species provide in the diet of the indigenous communities. Large mammals are defined as all species with a body weight greater than 10 kg (Emmons, 1997). The jaguar is included due to its ecological and spiritual importance as it is one of the strongest *Chimias* necessary for natural and spiritual balance
			Large frugivorous birds: Baudo guan (*Penelope ortoni*), yellow-throated toucan (*Ramphastos swainsonii*), collared aracari (*Pteroglossus sanguineus*), great curassow (*Crax rubra*)	Selected for two main reasons: (1) They populate a given area and perform fundamental ecological functions (Montaldo, 2005). These species maintain ecological interactions of mutualism with plants, as they feed on their fruits and disperse their seeds, thus guaranteeing the natural regeneration of the forest. With the disappearance of this group, these ecological interactions also disappear (Montaldo, 2005). (2) These are species of interest for the communities as food and pets
			Raw material for handicrafts: Mocora palm (Guerregue, *Astrocaryum standleyanum*), orange beads (Tetera, *Stromanthe lutea*), Chocolatillo (*Ischnosiphon arouma*), Puchama/Puchicama (*Arrabidaea chica*), Quitasol (*Mauritiella pacifica*)	Selected as a set of promising species that make the forest provide non-timber economic alternatives and provide income to indigenous communities. They are also closely related to ethnic cultural aspects of the *Wounaan-Nonam* ethnic group, and thus are under high pressure due to use

(continued)

Table 3.4 (continued)

"Guarantee the permanence of the natural environment or of some of its components, as a base for the maintenance of the cultural diversity of the country and the social appreciation of nature"	"Conserve natural spaces associated with material or immaterial cultural elements of ethnic groups"	Cultural	Spiritual elements: *Chimias, Dichardi,*[a] traditional medicine	Selected for two fundamental factors: (1) the close relationship between spirituality (cosmogony and cosmovision) of this culture and nature, and (2) they seek to rescue traditional elements of the culture that can show the road to reconnecting as an original people with nature and developing a new identity in accordance with the current needs and expectations of the communities, and thus generating guidelines to live harmoniously in the territory in an autonomous and environmentally sustainable manner
			Artistic elements: handicrafts, dances, corporal painting	Selected because there are a series of artistic practices such as the production of handicrafts, songs, dances, and body painting strongly associated with natural elements (flora and fauna species) that are part of the cultural identity of the *Wounaan-Nonam* ethnic group and are therefore the ancestral link between the community and natural resources
			Mother language *"Woun meu"*	Selected due to the close relationship between biological and linguistic diversity that undoubtedly allows understanding of the historical relationship that man has had with nature, given the domestication and use of diverse plants and animals present in these areas, which made it necessary to create new terms to explain nature. This allowed the construction of a directly proportional relationship between natural and linguistic diversity through generations. The loss of either of these components would result in the disappearance of the other (Toledo & Barrera, 2008)

[a] *Dichardi* is a sacred building where everyone meets and has the power of spirituality, in which knowledge is transmitted and the community is woven

non-human beings, and abiotic elements present in the environment. In this way, there is a systemic and integral vision regarding the relationship between wellness, health, and the environment. The wisdom of spiritual medics or shamans is called upon, who through their ancestral knowledge about nature and natural practices and rituals manage to heal diseases to restore a state of balance and harmony.

In this context, the territory is the space where culture, traditions, beliefs, feasts, and language can develop. Consequently, the environmental equilibrium of their territories is directly related to physical and mental health and to the development of their life plans. However, it is necessary to generate a correct articulation of the ancestral forms of traditional authority and the state, represented in the territorial entities (municipalities and governorate) and institutions, to consolidate a co-management strategy for the territory that eliminates or restricts the threats related to the presence of illegal armed groups and illicit activities.

Fig. 3.6 Examples of the biological and cultural values of conservation objects selected for the IPAs. (**a**) Large frugivorous birds: great curassow (*Crax rubra*); (**b**) large mammals: jaguar (*Panthera onca*); (**c**) fine timber: Chanul (*Humiriastrum procerum*); (**d**) raw material for handicrafts (Mocora palm (*Astrocaryum standleyanum*); (**e**) spiritual elements: *Dichardi* seen from the inside in a meeting with the community of Puerto Pizario; (**f**) artistic elements: Mocora palm handicrafts. (Photo credits: a, b, d by A. Quintero-Angel, 2016; c by A. Giraldo, 2017; and e, f by A. López-Rosada, 2016)

It is also important that the communities and the state ensure the implementation of ethno-development models in the *Wounaan-Nonam* territories, which would contribute to the survival of the people and, with them, the ancestral management and heirloom conservation practices. These practices have positive impacts on the health of the populations, from their intimate spiritual relationship with their environment to the use of biodiversity for the making of medicines, rituals, foodstuffs, and handicrafts, among others.

The ancestral link between the community and natural resources makes many species an important part of the traditional knowledge of the *Wounaan-Nonam* culture. Examples include medicinal plants and animal species used to strengthen human energy and maintain the spirits (*Chimias*) within the territory. Other examples are cultural practices such as handicrafts and body painting, which use natural inks. These communities recognise a decrease in the number of these species used in health and cultural practices. This favours the establishment of joint actions between the environmental authorities and the indigenous communities for biodiversity conservation. Consequently, it is essential to encourage the sustainable use of elements and raw materials associated with biodiversity, complemented by the implementation of enrichment programmes for species under greater pressure in the communities.

For the *Wounaan-Nonam* culture, all plants in nature are medicinal. The specific use of each plant and its information is handled exclusively by the traditional authority (i.e. spiritual medics or shamans). For this reason, this information could not be documented. Nonetheless, these communities recognise and use with frequency about 146 species of plants belonging to 47 families. Some examples of plants used to cure minor spiritual ailments, ward off evil *Chimias*, and attract good luck are palo de agua o nacedero (*Trichanthera gigantea*), Canilla de Venado (*Piper tricuspe*), Venezuelan pokeweed (*Phytolacca rivinoides*), neotropical snakefern (*Microgramma reptans*), and mouse tail (*Peperomia sp.*).

Even though the number of species used by the indigenous people is ample, it leaves out numerous species present in the region, catalogued as one of the richest and most diverse areas on the planet in terms of flora (Rangel et al., 2004). In addition, it is important to highlight that this knowledge of traditional plants is being lost due to the fact that new generations suffer from a loss of cultural identity, making it more difficult to pass on ancestral traditions. It is crucial to recover this ancestral medicinal knowledge as it has sociocultural implications and impacts the health of these communities, which in most cases only have traditional doctors for healthcare. Conventional health centres (i.e. Western medicine) are usually located several hours away by boat in towns or cities far away from the communities, and are too expensive to turn to.

The foodscape of the original people is also linked to the environment, and strengthened management of the territory, such as IPAs, contributes directly to the enhancement of their food hubs and nutritional security, along with improvement in health indicators. Although the dynamics of armed conflict hamper the free movement of community members through the environment, limiting food diversity, the sustainable use of the territory can be promoted so that internal regulations are established for use and exploitation of species associated with food security. Nevertheless, it is necessary to implement programmes to reinforce ancestral crops and cultural practices that contribute to the food sovereignty of these populations and to control the invasiveness of exotic species.

A key element for the conservation of the biocultural memory of the *Wounaan-Nonam* people is the conservation of the native language and ancestral cultural practices, which are also at risk of disappearing. Many of the words and names that the *Wounaan-Nonam* language uses to refer to elements associated with biodiversity could be at risk of disappearing or falling into disuse, due to factors such as isolation and state neglect that have undermined their cultural identity (Agencia de Noticias U. Nacional, 2014). Elements of their language are also used in practices associated with spiritual health in healing rituals. Likewise, these communities in recent years have experienced an accelerated process of acculturation, given the permanent interaction with other populations such as Afro-descendants and mestizos. Therefore, as stated in Sect. 3.2, the survival of ancestral cultural practices such as dance, handicrafts, or body painting is increasingly at risk. This highlights the urgent need for support of the indigenous education model so that the new generations perpetuate the use of the language in their daily life and appreciate and maintain cultural practices. In addition, the use of these artistic elements in spiritual and healing rituals contributes to the mental and spiritual well-being of these communities.

Given the increasing presence of illegal armed groups and forced interactions with the dominant society that accelerates the acculturation processes, a major effort to maintain the culture of the *Wounaan-Nonam* people is imperative. The fading of their biocultural memory has implications not only for the health and well-being of the indigenous population, but also for the degradation of ecosystems, which are in a good state of conservation within the IPAs, including the presence of highly threatened species of fauna and flora in the biogeographic Chocó region.

The survival of beliefs such as the *Chimias* has benefited conservation efforts as original people see nature as part of their spiritual world. This is evident in the case of timber species, which are conserved in indigenous territories because of their importance in spiritual terms, but are overexploited outside their territories because other communities (non-believers) only consider their commercial value (e.g. *Toxicodendron striatum*, or "aluvillo", and *Ceiba pentandra*, or "balsa"). The cosmovision of original peoples has, therefore, important and visible benefits for environmental health. However, a more decisive role on the part of the state and Colombian society is imperative, so that the processes of self-education and the conservation of biocultural memory may be favoured.

5 Conclusions and Lessons Learned

Historically, indigenous communities have had a mythological and spiritual relationship with natural resources, interacting with them not only as living beings, but also as spiritual entities, important for their survival and physical, mental, and spiritual well-being. Therefore, well-being is an integral part of the health of these original people. Accordingly, in situ conservation strategies, such as the IPAs, directly contribute to the livelihoods of these communities, improve management in their SEPLS, and promote conservation of their biocultural memory. IPAs have become a strategy to enrich the supply of indigenous families, because its biodiversity is directly related to the diets and eating habits, ancestral practices of traditional medicine, fashioning of handicrafts, and elements associated with spirituality and indigenous cosmogony.

The benefits of the efforts associated with biocultural memory conservation and the management of indigenous SEPLS that are shown in this case study have the potential to be replicated in the 115 original people communities in Colombia and in other ethnic communities around the world. The implemented actions of sustainable use of biodiversity and natural resources are reflected in the improvement of the fauna and flora populations (quality and quantity) inhabiting these areas, thus contributing to the nearby communities' quality of life (not only indigenous but also peasants, Afro-descendants, and recent mestizo settlers).

Consequently, IPAs serve as a conservation and resistance strategy, especially in ensuring the autonomy of the food hubs of the communities and a decrease in dependence on external inputs (i.e. food, supplies, goods), whose circulation is not allowed on many occasions due to the dynamics of the armed conflict. Likewise, by developing participatory inventories and exchanging knowledge on fauna and flora, traditional knowledge on foodstuff diversity is maintained.

In biocultural diversity conservation processes with indigenous communities, it is essential to assimilate the concepts of memory, symbolism, and myth, as they facilitate our understanding of their own knowledge by decoding their way of relating to natural resources. Only through the understanding of the relationship that indigenous women and men (i.e. gender perspective) have with their environment is any proposal for joint action for conservation of biocultural memory viable.

These issues, identified through this experience, are only the visible environmental phenomenon; they are the product of a nuanced matrix, or a set of interconnected cycling social problems such as forced displacement, cultural fragmentation, armed conflict, and poverty, among others. Only by reading the complexity of the context in the indigenous communities, or historicity, would it be possible to build effective participatory strategies to mitigate and/or solve specific conflicts associated with environmental health.

Although there is a legal framework recognising Colombia as a multiethnic and multicultural country, it is necessary to promote public practices of multiculturality, especially when they relate to indigenous SEPLS. Similarly, joint efforts for understanding state functioning and the logic of original people are needed in order to potentiate conservation strategies of biocultural memory and community governance.

Acknowledgements We are grateful to the communities in Puerto Pizario and Santa Rosa de Guayacán for their kindness and for allowing us to learn about their culture. Special thanks to their traditional authorities (*Cabildos*) who facilitated the logistics for our stay in these communities. This study was carried out within the framework of the agreement 011 CVC–ACIVA RP of 2013, entitled "Aunar esfuerzos técnicos económicos y humanos para adelantar las acciones necesarias orientadas a declarar áreas de interés ambiental en comunidades indígenas aliadas a la ACIVA-RP". The authors are also grateful to ACIVA-RP and CVC institutions for providing the funds for this study, to A. Giraldo for the donation to the indigenous community of the photograph of Chanul (*Humiriastrum procerum*) and the authorization for its use, and to the SITR vol. 7 Editorial Team for reviewing this manuscript.

References

Agencia de Noticias U. Nacional. (2014). Indígenas Wounaan, en riesgo de desaparecer, viewed 7 September 2021. https://www.elespectador.com/colombia/mas-regiones/indigenas-wounaan-en-riesgo-de-desaparecer-article-509707/.

Borrini-Feyerabend, G., Kothari, A., & Oviedo, G. (2004). Indigenous and local communities and protected areas: Towards equity and enhanced conservation. In *Guidance on policy and practice for co-managed protected areas and community conserved areas*. IUCN.

Burlingame, B., Charrondiere, R., Dernini, S., Stadlmayr, B., & Mondovì, S. (2012). Food biodiversity and sustainable diets: Implications of applications for food production and processing. In J. I. Boye & Y. Arcand (Eds.), *Green technologies in food production and processing* (Food engineering series) (pp. 643–658). Springer-Verlag. https://doi.org/10.1007/978-1-4614-1587-9_24

Cárdenas L.D. & Salinas, N.R. (eds.) 2007, Libro rojo de plantas de Colombia. Volumen 4. Especies maderables amenazadas: Primera parte, Instituto Amazónico de Investigaciones Científicas SINCHI—Ministerio de Ambiente, Vivienda y Desarrollo Territorial

Consejo Regional Indígena del Cauca (CRIC). (n.d.). *Guardia indígena*, viewed 1 February 2021. Retrieved from www.cric-colombia.org/portal/guardia-indigena/.

Consejo Superior de la Judicatura (ONIC). (2011). Módulo de Capacitación Intercultural indígena, Bogotá.

Corporación Autónoma Regional del Valle del Cauca (CVC) and Fundación Agua Viva (FUNAGUA). (2010). Aunar esfuerzos técnicos y económicos para realizar el análisis preliminar de la representatividad ecosistémica, a través de la recopilación, clasificación y ajuste de información primaria y secundaria con rectificaciones de campo del mapa de ecosistemas.

Corporación Autónoma Regional del Valle del Cauca (CVC). (2007). Construcción colectiva del sistema departamental de áreas protegidas del Valle del Cauca (SIDAP): Propuesta conceptual y metodológica.

Defensoría del Pueblo. (2018). Economías ilegales, actores armados y nuevos escenarios de riesgo en el posacuerdo, Bogotá.

Departamento Administrativo Nacional de Estadística (DANE). (2019). Población Indígena de Colombia, Bogotá.

Dudley, N., Higgins-Zogib, L., & Mansourian, S. (2009). The links between protected areas, faiths, and sacred natural sites. *Conservation Biology, 23*(3), 568–577. https://doi.org/10.1111/j.1523-1739.2009.01201.x

Ekblom, A., Shoemaker, A., Gillson, L., Lane, P., & Lindholm, K. J. (2019). Conservation through biocultural heritage—Examples from sub-Saharan Africa. *Land, 8*(1), 5.

Emmons, L. (1997). *Neotropical rainforest mammals: A field guide* (2nd ed.). The University of Chicago Press.

Flórez-Vargas, C. A. (2016). El concepto de derecho mayor: una aproximación, desde la cosmología andina. *DIXI, 18*(24). https://doi.org/10.16925/di.v18i24.1523

Garavito-Bermúdez, D. (2020). Biocultural learning—Beyond ecological knowledge transfer. *Journal of Environmental Planning and Management, 63*(10), 1791–1810. https://doi.org/10.1080/09640568.2019.1688651

Geilfus, F. (2002). *80 Herramientas Para El Desarrollo Participativo Diagnóstico, Planificación Monitoreo y Evaluación*. Instituto Interamericano de Cooperación para la Agricultura (IICA).

Granizo, T., Molina, M. E., Secaira, E., Herrera, B., Benítez, S., Maldonado, O., Libby, M., Arroyo, P., Ísola, S., & Castro, M. (2006). *Manual de Planificación para la Conservación de Áreas, PCA*. The Nature Conservancy & USAID.

Kawulich, B. (2005). La observación participante como método de recolección de datos. *Forum Qualitative Social Research*, **6**(2).

Lindholm, K., & Ekblom, A. (2019). A framework for exploring and managing biocultural heritage. *Anthropocene, 25*, 100195.

Lopez-Maldonado, Y., & Berkes, F. (2017). Restoring the environment, revitalizing the culture: Cenote conservation in Yucatan, Mexico. *Ecology and Society, 22*(4). https://doi.org/10.5751/ES-09648-220407.

Losos, E., & Leigh, E. G. (2004). *Tropical forest diversity and dynamism. Findings from a large-scale plot network*. University of Chicago Press.

Mamo, D. (2020). El mundo indígena. Documento internacional: Grupo de Trabajo Internacional para Asuntos Indígenas (IWIGIA).

Mejía, G. (2000). Presentación y Descripción Fonológica y Morfosintáctica del Waunana. In M. S. González de Pérez & M. L. Rodríguez de Montes (Eds.), *Lenguas indígenas de Colombia: Una visión descriptiva* (pp. 86–88). Instituto Caro y Cuervo, Santa fe de.

Ministerio de Cultura. (2009). *ABC del Plan Especial de Salvaguardia*, viewed 25 January 25 2021. Retrieved from https://www.mincultura.gov.co/areas/patrimonio/noticias/Paginas/2009-09-08_26664.aspx.

Ministerio de Cultura. (2010). Caracterización de los pueblos indígenas de Colombia. Wounaan tejedores de redes, Ministerio de Cultura.

Ministerio de Cultura. (2011). Auto-diagnóstico sociolingüistico-Wounaam.

Ministerio de Justicia and UNDOC. (2013). Caracterización regional de la problemática asociada a las drogas ilicitas en el departamento del Chocó.

Ministerio del Interior. (2012) Plan de Salvaguarda etnico del pueblo Wounaam de Colombia.

Montaldo, N. H. (2005). Aves frugívoras de un relicto de selva subtropical ribereña en Argentina: manipulación de frutos y destino de las semillas. *Hornero, 20*(2), 163–172.

Morales, J. C. (2012). La soberanía y autonomías alimentaris en Colombia. Revista Semillas, viewed 1 February 2021. Retrieved from http://www.semillas.org.co/es/resultado-busqueda/la-soberan.

Oldekop, J. A., Holmes, G., Harris, W. E., & Evans, K. L. (2016). A global assessment of the social and conservation outcomes of protected areas. *Conservation Biology, 30*(1), 133–141. https://doi.org/10.1111/cobi.12568

Organización Nacional Indígena de Colombia. (2018). *Pueblo Wounaam*, viewed 07 September 2021. Retrieved from https://www.onic.org.co/pueblos/1155-waunana.

Oviedo, G., & Jeanrenaud, S. (2007). Protecting sacred natural sites of indigenous and traditional peoples. In J. Mallarach, & T. Papayannis, (Eds.) *Protected areas and spirituality* (pp. 77–99). Proceedings of the first workshop of the Delos initiative, Montserrat 2006. IUCN, and Publicacions de l'Abadia de Montserrat, Gland, Switzerland, and Montserrat, Spain.

Plotkin, J. B., Potts, M. D., Yu, D. W., Bunyavejchewin, S., Condit, R., Foster, R., Hubbell, S., Lafrankie, J., Manokaran, N., Seng, L. H., Sukumar, R., Nowak, M. A., & Ashton, P. S. (2000). Predicting species diversity in tropical forests. *Proceedings of the National Academy of Sciences, 97*, 10850–10854.

PRATEC. (2012). *Diálogo de saberes: una aproximación epistemológica*. AMC Editores SAC.

Quintero-Ángel, A., López-Rosada, A., Miyela-Riascos, M., Tandioy-Chasoy, L. H., Gaitán-Naranjo, M. C., & Escobar-Sabogal, C. M. (2015). Recuperación de saberes e implementación de una herramienta pedagógica para la conservación de áreas protegidas indígenas del pueblo Nonam. In R. Monroy, J. García-Flores, M. Pino-Moreno, & C. N. Eraldo (Eds.), *S119 Saberes etnozoológicos Latinoamericanos* (pp. 57–72). UEFS Editora.

Quintero-Angel, M., Duque-Nivia, A. A., & Coles, A. (2020). A historical perspective of landscape appropriation and land use transitions in the Colombian South Pacific. *Ecological Economics, 181*, 106901. https://doi.org/10.1016/j.ecolecon.2020.106901

Rangel, J., Aguilar, M., Sanchez, H., & Lowy, P. (2004). Región Costa Pacífica. In J. Rangel (Ed.), *Colombia Diversidad Biótica I* (pp. 121–139). Instituto de Ciencias Naturales-Universidad Nacional de Colombia-Inderena.

Reyes-Ardila, J. A. (2019). Proceso de diseño estratégico en la creación de un modelo de negocio para la comunidad de mujeres indígenas Wounaan—Nonam en Bogotá, Universidad Ean.

Rodríguez-Sabiote, C., Llorente, T. P., & Gutiérrez-Pérez, J. (2006). La triangulación analítica como recurso para la validación de estudios de encuesta recurrentes e investigaciones de réplica en educación superior. *RELIEVE— Revista Electronica de Investigacion y Evaluacion Educativa, 12*(2), 289–305.

Sarmiento, F. O., & Viteri, X. (2015). Discursive heritage: Sustaining Andean cultural landscapes amidst environmental change. In K. Taylor, A. St Clair, & N. J. Mitchell (Eds.), *Conserving cultural landscapes: Challenges and new directions*. Routledge.

Sarmiento, F., & Hitchner, S. (Eds.). (2019). *Indigeneity and the sacred: Indigenous revival and the conservation of sacred natural sites in the Americas*. Berghahn Books.

Semana Sostenible. (2019). Pueblos indígenas, en el epicentro de los conflictos ambientales.

Speelman, E. N., Groot, J. C. J., García-Barrios, L. E., Kok, K., van Keulen, H., & Tittonell, P. (2014). From coping to adaptation to economic and institutional change—Trajectories of change in land-use management and social organization in a biosphere reserve community, Mexico. *Land Use Policy, 41*, 31–44. https://doi.org/10.1016/j.landusepol.2014.04.014

Toledo, V., & Barrera, N. (2008). *La memoria biocultural: la importancia ecológica de las sabidurías tradicionales*. Icaria.

Toledo, V., Barrera, N., & Boege, E. (2019). *Qué es el Diversidad Biocultural*. Universidad Nacional Autónoma de México.

United Nations. (2004). The concept of indigenous peoples: Background paper prepared by the secretariat of the permanent forum on indigenous issues, Document PFII/2004/WS.1/3, Department of Economic and Social Affairs, Workshop on Data Collection and Disaggregation for Indigenous Peoples, New York.

United Nations. (2007). The Declaration on the Rights of Indigenous Peoples. Resolution adopted by the General Assembly without reference to a Main Committee (A/61/L.67 and Add. 1).

Vélez, M. A., Robalino, J., Cardenas, J. C., Paz, A., & Pacay, E. (2020). Is collective titling enough to protect forests? Evidence from Afro-descendant communities in the Colombian Pacific Region. *World Development, 128*, 104837. https://doi.org/10.1016/j.worlddev.2019.104837

Walsh, C. (2008). Interculturalidad, plurinacionalidad y decolonialidad: las insurgencias político-epistémicas de refundar el Estado. *Tabura Rasa, 9*, 131–152.

Wild, R., & McLeod, C. (2008). *Sacred natural sites: Guidelines for protected area managers*. IUCN.

Laws and Regulations

Congress of Colombia, LAW 21 OF 1991Constitutional Court (2004). Decision T-025 of 2004.

Constitutional Court (2009). Auto 004 of 2009.

Decree 2164 of 1995 (1995). Official Diary No. 42.140 of 7 December 1995.

Instituto Colombiano de Reforma Agraria-INCORA (1983). Resolution No. 013 of 1983.

Instituto Colombiano de Reforma Agraria INCORA (1989). Resolution No. 054 of 1989.

The opinions expressed in this chapter are those of the author(s) and do not necessarily reflect the views of UNU-IAS, its Board of Directors, or the countries they represent.

Open Access This chapter is licenced under the terms of the Creative Commons Attribution 3.0 IGO Licence (http://creativecommons.org/licenses/by/3.0/igo/), which permits use, sharing, adaptation, distribution and reproduction in any medium or format, as long as you give appropriate credit to UNU-IAS, provide a link to the Creative Commons licence and indicate if changes were made.

The use of the UNU-IAS name and logo, shall be subject to a separate written licence agreement between UNU-IAS and the user and is not authorised as part of this CC BY 3.0 IGO licence. Note that the link provided above includes additional terms and conditions of the licence.

The images or other third party material in this chapter are included in the chapter's Creative Commons licence, unless indicated otherwise in a credit line to the material. If material is not included in the chapter's Creative Commons licence and your intended use is not permitted by statutory regulation or exceeds the permitted use, you will need to obtain permission directly from the copyright holder.

Chapter 4
SEPLS Well-Being as a Vision: Co-managing for Diversity, Connectivity, and Adaptive Capacity in Xinshe Village, Hualien County, Chinese Taipei

Paulina G. Karimova, Shao-Yu Yan, and Kuang-Chung Lee

Abstract Since 2016, a 600 hectare "ridge-to-reef" watershed of the Jialang River in Xinshe Village, Hualien County, Chinese Taipei, has been adaptively co-managed by a multi-stakeholder platform uniting two indigenous tribes (Fuxing and Xinshe) and four regional government agencies subordinate to the Council of Agriculture. The Five Perspectives of the Satoyama Initiative formed the core of the Xinshe SEPLS adaptive co-management model. The year 2020 marked the end of the short-term phase (2016–2019) and a transition period to the midterm phase (2021–2026) of the Xinshe "Forest-River-Village-Ocean" Eco-Agriculture Initiative (the Xinshe Initiative). How could the midterm management of the Xinshe Initiative most effectively enhance the Xinshe SEPLS well-being by 2026? To answer this question, we developed a set of 20 Localised Indicators of Resilience in the Xinshe SEPLS, analysed the concept of SEPLS well-being on the basis of the 5R conceptual framework ("ridge-to-reef", risks, resources, and resilience), and contributed the results of our study to the midterm action plan of the Xinshe Initiative.

Keywords SEPLS well-being · Adaptive co-management · 5R ("ridge-to-reef", risks, resources, and resilience) · Diversity · Connectivity · Adaptive capacity · Chinese Taipei

1 Introduction

1.1 Defining SEPLS Well-Being

"Societies in harmony with nature" is the overarching goal of the Satoyama Initiative and the 2050 vision reiterated in the zero draft of the Post-2020 Global Biodiversity

P. G. Karimova · S.-Y. Yan · K.-C. Lee (✉)
National Dong Hwa University, College of Environmental Studies, Hualien, Taiwan
e-mail: kclee@gms.ndhu.edu.tw

© The Author(s) 2022

M. Nishi et al. (eds.), *Biodiversity-Health-Sustainability Nexus in Socio-Ecological Production Landscapes and Seascapes (SEPLS)*, Satoyama Initiative Thematic Review, https://doi.org/10.1007/978-981-16-9893-4_4

Framework (Convention on Biological Diversity (CBD), 2020). The upcoming decades are expected to showcase diverse pathways towards achieving this vision at multiple scales, in different ecosystems and with a variety of stakeholders involved. Socio-ecological production landscapes and seascapes (SEPLS) have a high potential to provide invaluable and local-scale insights of their own (Nishi & Yamazaki, 2020).

One may wonder, however, what does harmony between societies and nature in SEPLS actually mean? Is it a win-win relationship, a balance, or the most desired state in SEPLS? Can it be put in concrete terms, observed, or even measured? To answer these questions, we propose the concept of *SEPLS well-being*.

The term "well-being" mainly derives from psychology and sociology studies where it is referred to as human well-being (Forgeard et al., 2011). It is generally defined as "the state of being or doing well in life; happy, healthy, or prosperous condition; moral or physical welfare (of a person or community)" (Oxford English Dictionary, 2021). Importantly, Dodge et al. (2012, p. 230) emphasise the non-static nature of well-being by calling it a "balance point between an individual's resource pool and challenges faced".

In a coupled socio-ecological system, like a SEPLS, we determine well-being to be a dynamic balance between the socio-economic and environmental *resources* available to a SEPLS and the socio-economic and environmental *risks* faced by it at a given point in time. As a SEPLS is subject to natural and human-induced, internal and external uncertainties, its good management should be capable of balancing out risks and resources in the most efficient way to realise, maintain, and enhance SEPLS well-being.

If the concept of SEPLS well-being can bring us one step closer to the vision of "societies in harmony with nature", it might be well worth to explore its practical on-the-ground application within a landscape approach. For this purpose, we examine Chinese Taipei's first ever experience with a multi-stakeholder platform for SEPLS revitalisation—the Xinshe "Forest-River-Village-Ocean" Eco-Agriculture Initiative (Xinshe Initiative).

1.2 Background: Xinshe Eco-Agriculture Initiative

The Xinshe SEPLS is a subtropical "ridge-to-reef" watershed of the Jialang River located in Xinshe Village, Fengbin Township, Hualien County, Chinese Taipei (Fig. 4.1a and Table 4.1). It covers an area of 600 hectares spanning from protected national forests of the Coastal Mountain Range, through production farmlands of two indigenous settlements (Amis Fuxing tribe (about 70 residents) and Kavalan Xinshe tribe (about 350 residents)) to the coral reef ecosystem of the Pacific Ocean (Lee et al., 2019, 2020) (Fig. 4.1b and Table 4.1). The landscape approach in the form of the Xinshe Initiative was introduced to the Xinshe SEPLS in October 2016 as a response to a number of socio-ecological challenges faced by the area over the years.

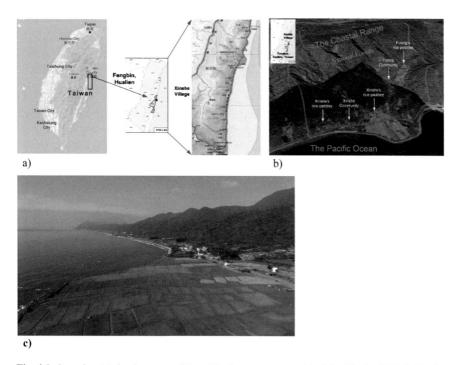

Fig. 4.1 Location (**a**), land-use map (**b**), and landscape-seascape (**c**) of the Xinshe SEPLS: Xinshe Village, Hualien County, Chinese Taipei (source: (**a**) and (**b**) Map data©Google, 2021, (**c**) photo taken by authors)

Table 4.1 Basic information of the study area

Country	Chinese Taipei
Province	Hualien County
District	Fengbin Township
Municipality	Xinshe Village
Size of geographical area (hectare)	1460
Number of direct beneficiaries (persons)	200
Number of indirect beneficiaries (persons)	665
Dominant ethnicity(ies), if appropriate	Indigenous Amis and Kavalan
Size of the case study/project area (hectare)	600
Geographic coordinates (latitude, longitude)	23°39′20.8″N, 121°32′21.8″E

Amis and Kavalan communities are widely known for their versatile skills as farmers, hunters, fishers, and gatherers who follow seasonal patterns and sustainably utilise the abundant resources of their surrounding environment all year round (Fig. 4.1c). Traditionally, the youth and the elderly, both men and women, all play indispensable roles in managing the SEPLS. Chinese Taipei's rapid industrial development in the 1970–1980s, however, brought with it the convenience of modern agriculture (including the introduction of chemical fertilisers, pesticides, and herbicides) and new socio-economic opportunities resulting in outmigration of youth to the cities. By the 2010s, depopulated and ageing, the Xinshe SEPLS was faced with deterioration of production farmlands, degradation of natural resources, loss of indigenous language and culture, and lack of incentives for the young people to return home.

The Xinshe Initiative (2016) became an effort to envision Xinshe SEPLS well-being by the means of SEPLS-wide promotion of eco-agriculture and uniting the sectoral efforts of three government agencies subordinate to the Council of Agriculture (COA) engaged in the area since 2010: Hualien Forest District Office (HFDOFB, since 2010), Hualien Branch of Soil and Water Conservation Bureau (HBSWC, since 2011), and Hualien District Agricultural Research and Extension Station (HDARES, since 2014).

Eco-agriculture in the context of the Xinshe Initiative is defined as "a fully integrated approach to agriculture, conservation and rural livelihoods" (Scherr & McNeely, 2008, p. 480) at the Xinshe landscape-seascape scale (Lee, 2016). The initial sectoral expertise of the three COA agencies working in the Xinshe SEPLS provided an opportunity to thoroughly address each of the three pillars of eco-agriculture: biodiversity (HFDOFB), production (HDARES), and livelihoods (HBSWC). For more information, please see Lee et al. (2019) in Volume 5 of the Satoyama Initiative Thematic Review.

The Xinshe Initiative further aligned the ecological, social, and economic aspects of eco-agriculture with the Five Perspectives of the Satoyama Initiative[1]: (a) ecosystem health and connectivity, (b) sustainable resource use, (c) traditions and innovation, (d) multi-stakeholder governance, and (e) sustainable livelihoods. These five eco-agricultural perspectives formed the thematic building blocks of the management cycle ("planning-implementation-evaluation-adjustment" stages) and the short-term action plan of the Xinshe Initiative (Fig. 4.2).

The Xinshe multi-stakeholder platform was established as a core mechanism for operationalising the Xinshe Initiative (Fig. 4.2). The main actors within the platform included primary stakeholders (representatives from the Fuxing and Xinshe tribes), secondary stakeholders (the four COA regional agencies) (the Eastern Region Branch of Agriculture and Food Agency (EBAFA) joined the above three in 2018 to support the "marketing" pillar), and the facilitator (National Dong Hwa University (NDHU—the authors)). Other relevant government agencies and tertiary stakeholders (non-governmental organisations, community organisations, and private enterprises) also partook in multi-stakeholder platform meetings when necessary (Lee et al., 2019).

From the very beginning, the Xinshe Initiative was projected as a decade-long revitalisation effort to be carried out in short-term (3 years) and midterm (7 years)

---

[1]For the Xinshe Initiative, the initial Five Perspectives of the Three-Fold Approach for the Satoyama Initiative were slightly modified from their original version (IPSI Secretariat, 2014, p. 9): (a) resource use within the carrying capacity and resilience of the environment, (b) cyclic use of natural resources, (c) recognition of the value and importance of local traditions and cultures, (d) multi-stakeholder participation and collaboration, and (e) contributions to socio-economies.

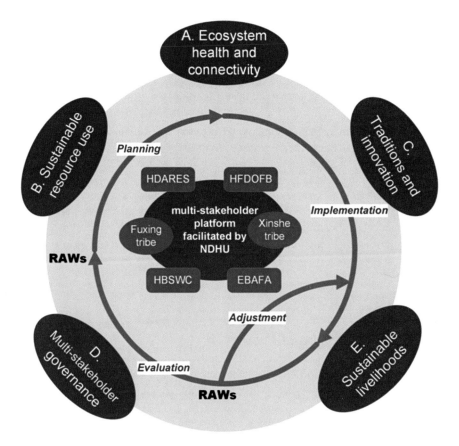

Fig. 4.2 The Xinshe "Forest-River-Village-Ocean" Eco-Agriculture Initiative and its adaptive co-management model based on the Five Perspectives of the Satoyama Initiative (source: prepared by authors). *HDARES* Hualien District Agricultural Research and Extension Station, *HFDOFB* Hualien Forest District Office, *HBSWC* Hualien Branch of Soil and Water Conservation Bureau, *EBAFA* Eastern Region Branch of Agriculture and Food Agency, *NDHU* National Dong Hwa University, *RAWs* resilience assessment workshops

phases. Ten years were deemed as a realistic enough period of time to allow the local elderly to observe the positive trends towards enhancement of Xinshe SEPLS well-being, and for the youth to return home to new opportunities and to take over the SEPLS' management in the long term.

1.3   Rationale: The Xinshe SEPLS Well-Being as a Midterm Vision

During the initial 3 years of the Xinshe Initiative, the multi-stakeholder partnership was characterised by several notable features. Firstly, it was a time of learning to work together in the cross-sectoral (between government agencies), cross-settlement (between Fuxing and Xinshe tribes), and cross-knowledge (expert and local) dimensions (Lee et al., 2019). Secondly, it allowed for "harvesting the low-hanging fruit"—accomplishing the most urgent action tasks based on the immediate challenges faced by the communities. Thirdly, the short-term phase laid out the "stepping stones" for issues to be addressed in the midterm phase (Karimova, 2021). In sum, these were the foundation years towards achieving Xinshe SEPLS well-being.

The year 2020 marked the end of the short-term phase (2016–2019) and served as a transition period between the phases. Evaluation of the short-term phase and planning for the midterm phase (2021–2026) were carried out at this time. At this pivotal stage of the Xinshe Initiative, multiple stakeholders pondered on a question: *How can the midterm management of the Xinshe Initiative most effectively enhance Xinshe SEPLS well-being by 2026?* Our research team saw this as an opportunity to give a practical application to the concept of SEPLS well-being and to contribute the results of our study to the midterm action plan.

2 Methods

2.1 The 5R Conceptual Framework

Our analysis of the Xinshe SEPLS well-being integrated the familiar concepts of "ridge-to-reef" (R-to-R), risks and resources (2R), and resilience into a 5R conceptual framework (Fig. 4.3).

We adapted Dodge et al.'s (2012, p. 230) model where well-being is positioned at the centre of a see-saw as "the balance point between an individual's resource pool and the challenges faced". Internal and external socio-economic and environmental threats, pressures, and challenges of SEPLS are placed into the *Risks* box, while the strengths, opportunities, and skills of all SEPLS elements (human and natural) are put into the *Resources* box.

The see-saw itself is envisioned as *Resilience*—a dynamic process in which a SEPLS utilises its available resources in order to adapt and grow in the face of risks,

Fig. 4.3 The 5R conceptual framework (adapted from Dodge et al., 2012)

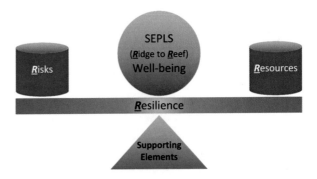

yet retaining its essential structure and functions (Brown, 2016; Ford et al., 2020; Carpenter et al., 2001). Having identified this fifth "R" in our conceptual framework, we employed community-based resilience assessment workshops (RAWs) as a tool (Lee et al., 2020; Fig. 4.2). Also, pictured as a supporting frame of the see-saw (lower pyramid structure in Fig. 4.3), our conceptual framework supposes that certain supporting elements are required in order to fulfil the 5Rs.

Therefore, we outlined the following objectives for our analysis: to elicit the socio-economic and environmental risks and resources of the Xinshe SEPLS at the end of the short-term phase, and to identify the supporting elements for propelling Xinshe SEPLS well-being in the midterm phase.

2.2 Resilience Assessment Workshops

We define resilience assessment workshops (RAWs) as a series of community-based activities aimed at the evaluation of socio-ecological resilience in a SEPLS, organised for the purpose of providing a problem-oriented feedback to the "adjust-ment" or "planning" stages of its adaptive co-management cycle (Lee et al., 2020; Fig. 4.2).

RAWs were first carried out in the Xinshe SEPLS in 2017–2018 during the "evaluation-adjustment" stages of the short-term phase. At the time, the original set of 20 Indicators of Resilience in SEPLS (Bergamini et al., 2014) was directly translated into the Chinese language and applied in RAWs (Lee et al., 2020). Mirroring the Five Perspectives of the Satoyama Initiative, the 2017–2018 RAWs played a crucial role in yielding community-based adjustments to the short-term action plan and received positive acclaim from both the Fuxing and Xinshe com-munities and the government agencies. This experience resulted in RAWs becoming the official "evaluation-adjustment-planning" tool of the Xinshe Initiative.

In 2020, however, RAWs were entrusted with the more complex task of evalu-ation and planning for the midterm phase (Fig. 4.2). It meant that at this stage the indicators for measuring resilience had to be *comprehensive* enough to reflect the multitude of challenges and opportunities in the Xinshe SEPLS, *relevant* enough to the local context and the action tasks of the Xinshe Initiative, and *comprehensible* enough to be easily understood by the local communities.

For this reason, the process of conducting the 2020 RAWs consisted of three consecutive stages: pre-RAWs (developing a set of 20 Localised Indicators of Resilience in the Xinshe SEPLS—Appendix A), core RAWs (conducting a series of 12 RAWs in the Fuxing and Xinshe communities), and post-RAWs (communi-cating assessment results to the multiple stakeholders and producing a bottom-up midterm action plan for the Xinshe Initiative). A detailed description of activities during each stage is presented in Appendix B and an evaluation sheet sample of Perspective A (Indicators A1–A4) in Appendix C. Please see Sun et al. (2020) for the photographs of the 2020 RAWs.

2.3 Resilience Assessment as the Xinshe SEPLS Health Check

As RAWs are a *community-based* tool for assessing resilience in SEPLS, it is imperative that the local people fully understand the purpose of this exercise, its benefits for the community, and, first and foremost, the meaning of what is being measured—resilience itself. The feedback from the Fuxing and Xinshe communities after a series of initial RAWs in 2017–2018 (Lee et al., 2020) and opinions voiced by government representatives during the 2020 pre-RAWs' preparatory workshop (Appendix B) warned us that *comprehensibility* was key. While the purpose and the benefits of RAWs were clear, the concept of resilience remained rather vague and was deemed "too academic" for both the local people and the government officials. How could we explain it better?

For the 2020 RAWs, we found a solution by not only providing a definition of resilience in "risks-resources" terms but also using the analogy of a human body and human health—something that all participants were most familiar with (Fig. 4.4). We started by referring to a healthy human as one who does not get sick easily and who recovers quickly even when some ailment occurs. We noted that the same could be said about a resilient SEPLS—it is less vulnerable to risks and has a strong capacity to adapt (find resources to cope with risks). Then we explained that similar to a human body being comprised of multiple organs which form its ten main systems, SEPLS is also home to numerous multifunctional natural and man-made elements.

In the same way that a doctor performs a health check (e.g. blood pressure, blood tests, temperature check) to find out the functionality of systems and organs within a

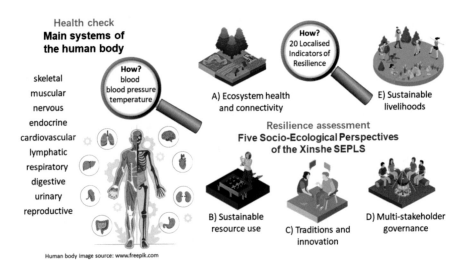

Fig. 4.4 Resilience assessment as the Xinshe SEPLS health check analogy (source: prepared by authors, with the human body image elements adapted from www.freepik.com (2021))

human body, a resilience assessment measures risks and resources within the Five Perspectives of a SEPLS with the help of the 20 Localised Indicators of Resilience as a measurement tool. We further emphasised that resembling the interconnectedness between the main systems in a human body, SEPLS elements also form a nexus pattern of their own. Thus, we encouraged participants in RAWs to be mindful of how risks and resources within one perspective could be linked to risks and resources within another.

3 Results and Discussion

3.1 Risks and Resources of the Xinshe SEPLS

Design of each perspective-related evaluation sheet (Appendix C) implied two types of assessment results: quantitative (scoring of indicators based on 1 star "very low" to 5 stars "very high" (Fig. 4.5)) and qualitative (discussion of risks and resources based on the participants' scoring decisions and priority ranking of indicator-relevant local examples of specific action tasks (Fig. 4.6)).

In the following Sects. 3.1 and 3.2, we make a particular use of the qualitative assessment results to discuss the socio-economic and environmental risks and resources within each of the five perspectives (Fig. 4.6, upper part) and elicit the supporting elements for the Xinshe SEPLS well-being (Fig. 4.6, bottom part). In Sect. 3.3, we come back to Fig. 4.5 by showing how the scoring results for the 20 Localised Indicators guided the choice of ten thematic priority areas for the midterm action plan of the Xinshe Initiative.

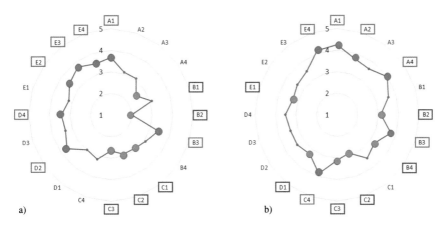

Fig. 4.5 Scoring results for 20 Localised Indicators of Resilience in the Xinshe SEPLS: (**a**) Fuxing tribe, (**b**) Xinshe tribe. Marked in green—indicators with (relatively) highest scores (3.42–4.25), in red—indicators with (relatively) lowest scores (2–2.85) (source: prepared by authors)

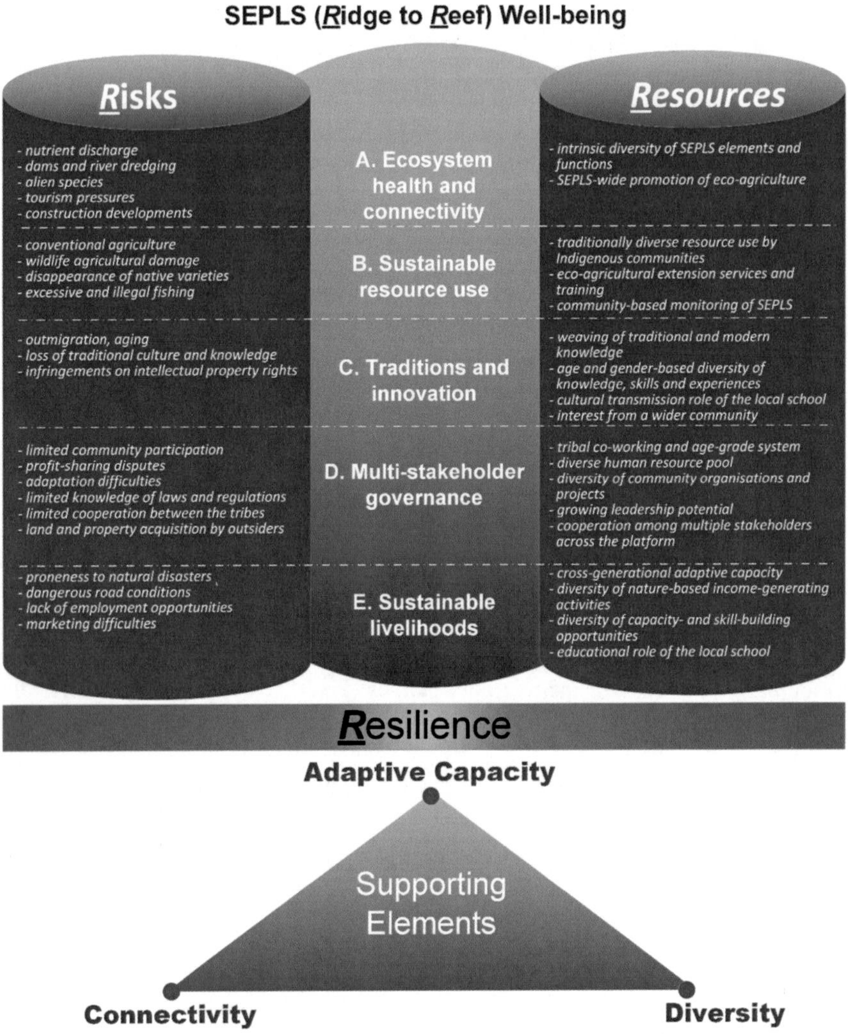

SEPLS (Ridge to Reef) Well-being

Risks		Resources
- nutrient discharge - dams and river dredging - alien species - tourism pressures - construction developments	**A. Ecosystem health and connectivity**	- intrinsic diversity of SEPLS elements and functions - SEPLS-wide promotion of eco-agriculture
- conventional agriculture - wildlife agricultural damage - disappearance of native varieties - excessive and illegal fishing	**B. Sustainable resource use**	- traditionally diverse resource use by Indigenous communities - eco-agricultural extension services and training - community-based monitoring of SEPLS
- outmigration, aging - loss of traditional culture and knowledge - infringements on intellectual property rights	**C. Traditions and innovation**	- weaving of traditional and modern knowledge - age and gender-based diversity of knowledge, skills and experiences - cultural transmission role of the local school - interest from a wider community
- limited community participation - profit-sharing disputes - adaptation difficulties - limited knowledge of laws and regulations - limited cooperation between the tribes - land and property acquisition by outsiders	**D. Multi-stakeholder governance**	- tribal co-working and age-grade system - diverse human resource pool - diversity of community organisations and projects - growing leadership potential - cooperation among multiple stakeholders across the platform
- proneness to natural disasters - dangerous road conditions - lack of employment opportunities - marketing difficulties	**E. Sustainable livelihoods**	- cross-generational adaptive capacity - diversity of nature-based income-generating activities - diversity of capacity- and skill-building opportunities - educational role of the local school

Resilience

Adaptive Capacity

Supporting Elements

Connectivity   **Diversity**

Fig. 4.6 The 5R of the Xinshe SEPLS well-being (source: prepared by authors)

Ecosystem Health and Connectivity

The "ridge-to-reef" natural environment of the Xinshe SEPLS is subject to a number of environmental and human-induced stressors. Heavy rains brought by seasonal monsoons (January–March and May–June) and typhoons (July–September) create a risk of floods and landslides (in the Fuxing area in particular). In the 1990s, in order to minimise the risk of natural disasters to community livelihoods (Perspective E),[2] a series of dams were constructed on the Jialang River. Together with excavator dredging of the river basin, these engineering constructions have become a serious obstacle for the spawning migration of fish and shrimp (21 rare varieties).

To date, conventional agricultural practices (Perspective B) are among the main human-induced risks to Xinshe ecosystem health. Given the undoubtable convenience of synthetic fertilisers and herbicides, physical constraints of the aged population, and a lack of young labour force, promotion of eco-agricultural production in the Xinshe SEPLS is a complex task. High nutrient concentration rates (N and P) from fertiliser overuse (mainly in the Xinshe community) can adversely impact the quality of drinking and irrigation water, which in turn can put stress on the coral reef ecosystem downstream.

In addition to the socio-economic risks, the scenic beauty of the Xinshe landscape-seascape has become a strong attractor to a growing number of tourists driving along the Coastal Highway #11. In early to mid-2020, following the international travel restrictions imposed by the Chinese Taipei Government in response to COVID-19 (19 March 2020), domestic tourism rose substantially (Karimova & Lee, 2021). As a result, littering, trampling on the paddy fields, and noise pollution became common phenomena in the Xinshe SEPLS. The residents also expressed their concerns about the risks of landscape fragmentation and threats to eco-agricultural production caused by such recent "construction invasions" as utility poles, high-voltage streetlights, and solar panels.

Having listed the above threats to SEPLS ecosystem health, the Fuxing and Xinshe communities, however, noted the intrinsic diversity of SEPLS elements and their functions as an invaluable resource. To them, floral and faunal diversity of the surrounding forests, aquatic and biotic resources of the Jialang River, agrobiodiversity of eco-friendly farmlands, and marine resources of the Pacific coastal zone can provide a strong buffer to natural and man-made threats. Also, participants in RAWs highlighted the imperative role of SEPLS-wide promotion of eco-agriculture during the short-term phase of the Xinshe Initiative in fostering the "forest-river-village-ocean" connectivity across the SEPLS. In their opinion, positive changes are already visible.

Sustainable Resource Use

Discussion of sustainable resource use in the Xinshe SEPLS was focused on common natural resources (forest, river, and ocean) and agricultural farmlands.

Due to the legally protected status of the national forests, their risks seemed to be of the least concern. Meanwhile, both communities pointed out unsustainable fishing

---

[2] In many cases, a risk or a resource within one Perspective was closely correlated to a risk or a resource within another Perspective. We demonstrate these nexus relationships by listing other relevant Perspectives in parentheses (here: Perspective E).

practices, including overfishing and intrusions from outsiders, as the highest-ranking resource-related risk in the Xinshe SEPLS. Electric fishing in the Jialang River and the trawl nets of commercial vessels in the nearby Pacific coastal zone have made Fuxing and Xinshe residents not only concerned about the fish stocks and ecosystem health (Perspective A), but also question traditional land rights of the communities (Perspective D).

In addition to looming threats from conventional agriculture, eco-agricultural production in Xinshe SEPLS is at risk from disturbances caused by wildlife (Formosan macaques, wild boar, wild hare, and barking deer). In the Fuxing community, for instance, where wildlife can easily roam into the production farmlands, the farmers' crop diversity is substantially limited. They have given up on growing any sweet vegetables (sweet potato, corn, or beetroot) as those are particularly targeted by the wildlife. Moreover, disappearance of native plant and animal varieties is also a big concern especially in the face of an increased risk of alien species' encroachment.

A well-established community-based monitoring of local resources (forest patrol, river brigades, and coral reef check-up teams) was listed by the RAWs' participants as the most immediate counterweight to the resource-use risks in the SEPLS. The traditionally diverse and sustainable ecosystem-based production activities of the indigenous communities (farming, hunting, fishing, arts and crafts, weaving, and culinary art) were highlighted as well (Perspective C). Improvement of eco-agricultural farming practices with the help of extension services and training provided by government agencies (HDARES and HFDOFB) was noted to be a valuable resource to tackle risks in production farmlands.

Traditions and Innovation

Outmigration, ageing, lack of young people, and, as a result, limited human and temporal capacities for documentation and transfer of traditional culture and knowledge pose by far one of the biggest risks for the Xinshe SEPLS. Disappearance of the indigenous Amis and Kavalan languages and loss of knowledge and practical skills of the elders are greatly feared by the locals. Return of out-migrated local youth to their homeland was the main hope voiced by almost everyone throughout RAWs (in this context as well as for all other perspectives).

In the meanwhile, since the introduction of the Xinshe Initiative, the locals have gained more hope in the future of their traditions and culture. They noted, for example, the important role that the weaving of traditional and modern knowledge plays in addressing such SEPLS issues as pest and wildlife management, reintroduction of native species and local seed varieties, and improvement of planting technologies (Perspective B). Diversity of knowledge, skills, and experiences of the elderly and youth, men and women, Kavalan, Amis, and Bunun[3] indigenous community members within the SEPLS is also highly valued by the locals. Particularly worth noting is the influential role of the Xinshe Primary School—the only educational institution in the Xinshe Village—as the centre of cultural and community cohesion.

Though local resources for cultural documentation and transmission are rather limited, there is a growing interest from research and media communities attracted to the Xinshe SEPLS either by the uniqueness of its indigenous culture and its scenic beauty or by the Xinshe Initiative itself—Chinese Taipei's one-of-a-kind multistakeholder initiative. Remarkably, this interest is both a risk and a resource for the SEPLS. On the one hand, if ethical considerations are not properly addressed, intellectual property rights to folklore and language resources end up lost to outsiders and the communities receive no benefit. On the other hand, it can be seen as an advantage: trained professionals can assist in documentation and promotion of traditional culture and knowledge as well as spur public interest and support towards the Xinshe SEPLS.

Multi-stakeholder Governance

In the Xinshe SEPLS, we view multi-stakeholder governance as a "double-layered" phenomenon, which includes the social capital of the Fuxing and Xinshe communities (the "inner layer") and the cooperation of multiple stakeholders across the platform (the "outer layer").

On the "inner layer", RAWs' participants have mentioned a limited participation of community members in SEPLS revitalisation activities (including the Xinshe multi-stakeholder platform itself) which stems from their unawareness, lack or interest, or distrust. Participation in community development projects (even those related to the Xinshe Initiative) is often at risk of benefit-sharing disputes within the communities. Also, despite the urge for young people to return to the SEPLS, they are often faced with the difficulties of cultural (language, local traditions, and customs), social (participation in local institutions), and economic (search for jobs) adaptation. This can lead to further misunderstandings and conflicts. In recent years, the threat of selling land and property to outsiders is a direct result of lack of prior communication among community members.

The "outer" layer of Xinshe multi-stakeholder governance is mainly at risk of water-related disputes between the Fuxing and Xinshe tribes related to the upstream-downstream management of the Jialang River's water quantity and quality (Perspective B). Until now, most concerns have been addressed on a person-to-person rather than a community-to-community basis, while many participants from the Xinshe community voiced the need for the latter.

---

[3] The third, less populous, indigenous group in the Xinshe SEPLS.

One of the strongest social capital resources in the Fuxing and Xinshe tribes is *Mipaliw*—a traditional tribal co-working and cooperation mechanism of "neighbour helping neighbour". The tribal age-grade system (in the Xinshe community) is another source of cohesion. Diversity of knowledge, skills, and experiences of community members within and outside of the SEPLS (including the youth residing in the cities) ensures a human resource pool.

Diversity of organisations within the communities (tribal councils, community development committees, indigenous language promotion centres, art workshops, etc.) and community projects supported by government agencies, private enterprises, and NGOs are resources for enhanced social inclusiveness, skill-building, and employment opportunities (Perspective E). Also, over the short-term years, the Xinshe multi-stakeholder platform has encouraged dialogue between young leaders from the Fuxing and Xinshe tribes and an enthusiastic collaboration across the platform on a number of SEPLS issues.

Sustainable Livelihoods

As mentioned in Perspective A, the proneness of the Xinshe SEPLS to natural disasters (typhoons, landslides, and floods) endangers community safety with a particular vulnerability of the elderly. Road conditions and traffic safety are also a matter of concern. Frequency of traffic accidents significantly increases during the rainy seasons due to the steep slippery road conditions in the Fuxing community, and during the holiday seasons on the busy Coastal Highway #11 passing through the Xinshe community.

Other livelihood risks are related to a lack of employment opportunities that prevents young people from returning to the SEPLS (Perspective D). A long-term pursuit of eco-agricultural production (Perspective B) as an income-generating activity may be hampered by marketing obstacles such as limited access to processing technology and equipment, as well as difficulties with organic and environmentally friendly certification, and a lack of marketing channels and skills.

A cross-generational adaptive capacity (past experiences of the local elders, Perspective C) to natural and socio-economic calamities is surely one of the most valuable livelihood resources in the SEPLS. Potential diversity of ecosystem-based income-generating activities (Perspective B) such as eco-agriculture, fishing, farming, eco- (marine) tourism, environmental interpretation, and arts and crafts can also generate a pool of employment opportunities. In the meanwhile, the good reputation of the Xinshe Primary School as one of Chinese Taipei's exemplary rural schools creates a strong incentive for the young people and their families to return to the Xinshe SEPLS.

An already evident connectivity between eco-agricultural production and human (physical, mental, spiritual) and ecosystem health creates a strong stimulus for proceeding with environmentally friendly production activities. Also, a gradual emergence of community-based local product brands (rice, tea, and natural dye) and experimenting with various marketing channels (weekend markets, indigenous product exhibitions, and online platforms)—especially at times of COVID-19—all contribute to fostering the marketing capacity of the Xinshe SEPLS.

3.2 Resilience: Supporting Elements for the Xinshe SEPLS Well-Being

With the risks and resources of the Xinshe SEPLS identified, we return to the 5R conceptual framework (Fig. 4.3) to take a closer look at the supporting elements of the see-saw structure of SEPLS well-being. Our analysis revealed three of them: diversity, connectivity, and adaptive capacity (Fig. 4.6, bottom).

Diversity is the basic supporting element. It represents diversity of all kinds and across all socio-economic and ecological risks and resources of the Xinshe SEPLS: of ecosystems and their functions (forest, freshwater, farmland, and marine), of wild and cultivated floral and faunal varieties (biodiversity, including agro-biodiversity), of soil micro- and macro-organisms (soil biodiversity), of resource use and production activities of the local people (farming, hunting, fishing, gathering, ecotourism, arts and crafts, and culinary art), of knowledge types (traditional and modern) and knowledge holders (the elderly and the young, men and women), of skills and expertise within and outside the SEPLS (including the youth residing in the cities), of capacity-building and partnership opportunities, of ecosystem-based income-generating activities, of diverse marketing channels (weekend markets, indigenous product exhibitions, and online marketing), and of past adaptation experiences and coping strategies (cross-generational adaptive knowledge).

Connectivity is the second supporting element that links the "many diversities" of SEPLS in a way that keeps them functional. Connectivity implies the nexus patterns *within* risks and resources (i.e. how occurrence of one risk may lead to another or how mobilisation of one resource may encourage mobilisation of another) and also *between* them (addressing risks with resources and the "spillover effect" on other risk-resource interactions). It ensures a reciprocal relationship and consideration of a "cause-effect" interdependence between all elements in the SEPLS in multiple dimensions: natural-to-natural ("ridge-to-reef") (forest-river-farmlands (as an ecosystem)-ocean), human-to-human (within the settlements, between the settlements, and across the multi-stakeholder platform), and human-to-natural (all aspects of sustainable resource use). With connectivity in mind, for instance, the link between fertiliser overuse, coral reef bleaching, diminishing fish stocks, and restrained fish spawning migration becomes evident. This should naturally lead to more sustainable production methods such as a shift to eco-agriculture.

Adaptive capacity is the third supporting element of Xinshe SEPLS well-being. We see it as the ability to mobilise diverse socio-economic and ecological resources and connect them in the most suitable way in order to face socio-economic and ecological risks. Adaptive capacity is the dynamic element that propels the see-saw of Xinshe SEPLS well-being forward and enables it to change and grow over time while maintaining its essential structure and functions. We note that along with diversity and connectivity, it is adaptive capacity that supports the resilience of the Xinshe SEPLS.

3.3 Adaptive Co-management for the Xinshe SEPLS Well-Being in the Midterm Phase (2021–2026)

Now we come back to the question asked by the multiple stakeholders in early 2020: *How can the midterm management of the Xinshe Initiative most effectively enhance Xinshe SEPLS well-being by 2026?* Answering this question was an essential part of the post-RAWs' analysis and planning process (Appendix B).

Indicators with the lowest scores (marked in red in Fig. 4.5) pointed out the aspects of SEPLS management that were of the highest concern to the local communities and required further improvement. These included B4 and D1—sustainable use and traditional rights in relation to common resources (river and ocean); B1, B2, and C3—self-sufficiency and appropriate technology in eco-agricultural production, and conservation and reintroduction of native species; C1 and C2—transfer and documentation of traditional ecological knowledge (TEK) and return of migrant youth; and E1 and E3—traffic safety, elderly-friendly infrastructure, and marketing capacity. Multi-stakeholder governance (Perspective D) was emphasised by almost all RAWs' participants as an important mechanism to maintain and enhance partnership for common objectives.

Thus, based on the scoring results (Fig. 4.5) and an in-depth discussion of environmental and socio-economic risks and resources in the Xinshe SEPLS (Fig. 4.6), we were able to outline ten thematic priority areas for the midterm phase of the Xinshe Initiative—two problem-oriented themes for each of the five perspectives (Fig. 4.7). To stipulate division of responsibilities among the multiple stakeholders in relation to the previously existing and newly emerged action tasks, we further combined the ten themes into the "4 + 2 + 1 Model" (three colours in Fig. 4.7: "4" red, "2" blue, and "1" green).

Four thematic areas in red colour ("4") are to be guided by the four regional government agencies within the COA system (HDARES, HFDOFB, HBSWC, and EBAFA) with each agency being primarily responsible for one of the themes. These priority areas include protection and sustainable use of aquatic and biotic resources of the Jialang River system (HFDOFB), the entire Perspective B—promotion of eco-agriculture and reintroduction of native agricultural varieties (HDARES), building up marketing channels and skills (EBAFA), and disaster risk prevention and safe and eco-friendly infrastructure (HBSWC).

Two themes in blue colour ("2") are the topics outside of the area of expertise of the COA and its subordinate agencies. They, therefore, require efforts from other relevant government agencies, NGOs, and/or private enterprises. One of the topics is related to protection and sustainable use of marine resources, where assistance from

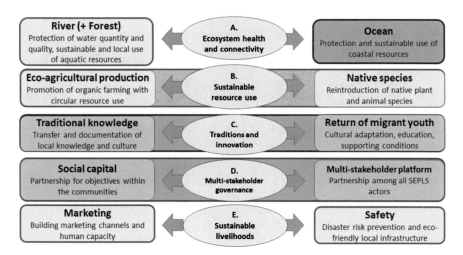

Fig. 4.7 The "4 + 2 + 1 Model" and ten thematic priority areas for midterm adaptive co-management in the Xinshe SEPLS (source: prepared by authors)

Hualien County Government, Taiwan Fisheries Research Institute, and ecological NGOs would be particularly valuable. The other is focused fully on Perspective C: the documentation and transfer of TEK and the return of migrant youth to the Xinshe SEPLS. Potentially, a number of government projects (by HFDOFB, HBSWC, and Taiwan Council of Indigenous People) could support the Fuxing and Xinshe communities and the Xinshe Primary School in addressing this theme.

The green box ("1") is comprised of the "inner" and "outer" layers of Xinshe multi-stakeholder governance. It is a cooperation and partnership mechanism of the midterm phase that can enable realisation of the "4 + 2" themes by the means of enhanced social capital within the communities and cooperation across the multi-stakeholder platform for the Xinshe Initiative.

Importantly, the interconnectedness of risks and resources within and between the Five Perspectives of the Xinshe SEPLS (see Sect. 3.1) is evident in the nexus patterns between the ten thematic priority areas of the "4 + 2 + 1 Model" as well. For instance, promotion of eco-agriculture and reintroduction of native agricultural varieties (B) will largely depend on the inclusion of TEK (C). In turn, eco-agricultural benefits are likely to have a far-reaching effect not only on ecosystem health (the Jialang River water quality upstream and downstream, restored coral reef ecosystem of the Pacific coast (A)) but also on human physical and mental health (E). Good environment and availability of income-generating opportunities (E) created by eco-agricultural production (B) can act as a strong incentive for the migrant youth to return to the Xinshe SEPLS (C). Diversity of skills, knowledge, and backgrounds of local and newly returned community members can enhance social capital (D) and provide new solutions to some urgent issues (e.g. open up a variety of marketing channels (E)).

Adaptive by design and in pursuit of Xinshe SEPLS well-being, the midterm action plan for the Xinshe Initiative (Appendix D) recommends the "4 + 2 + 1 Model" and ten thematic priority areas to remain at focus throughout the entire midterm phase (up to 2026). The action plan also provides a checklist of 44 specific action tasks—the bottom-up concerns and suggestions of the 2020 RAWs' participants. The checklist is meant to serve as a helpful action guidance to all multiple stakeholders during the 3-year (2021–2023) period and is to be adjusted at the end of each year. A comprehensive reassessment of the action tasks is scheduled to be performed in "evaluation-adjustment" follow-up RAWs in 2023.

4 Lessons Learned and Conclusions

Stretching from ridge to reef, both a landscape and a seascape, blessed with natural beauty and the wisdom of its indigenous inhabitants, the Xinshe SEPLS is truly a gem of Chinese Taipei's east coast. Well known across the island as an attractive tourist destination, in recent years, Xinshe has been gaining increasing attention for its revitalisation efforts inspired by the Satoyama Initiative. The Xinshe Initiative and its multi-stakeholder platform have already become a knowledge-sharing base for *Satoyama* practitioners across Chinese Taipei. The findings presented in this study add to this experience exchange from the perspective of SEPLS well-being.

Since 2016, promotion of eco-agriculture has played a major role in *enhancing various aspects of human health*: physical (green and nutritious produce from clean soil and water, reduced proneness to seasonal diseases, enhanced self-sufficiency at times of COVID-19), psychological and mental health (sense of place, sense of unity with natural surroundings), cultural and spiritual health (revival of TEK such as shaman rituals, harvest, and ocean festivals), and community health (cohesion, sense of comradeship, and return of migrant youth).

A number of *synergies and trade-offs* have also been observed. For example, 4 years (2016–2020) of eco-agricultural practices on land have already led to the improved health of the coral reef ecosystem (more fish, more coral diversity, and less coral bleaching)—a type of landscape-seascape "ridge-to-reef" synergy. On the other hand, the midterm management should be particularly mindful of the socio-economic trade-offs in Xinshe SEPLS dynamics. For instance, the much-anticipated return of the migrant youth (an actively ongoing process during COVID-19) can have a positive effect on the diversity of community organisations and projects but can also result in adaptation difficulties, conflicting objectives, and profit-sharing disputes between old and new community members.

We found the 20 Localised Indicators of Resilience to be a useful tool in *measuring the effectiveness of Xinshe SEPLS management*. By being comprehensive, they addressed a wide spectrum of socio-economic and environmental risks and resources in the SEPLS. Likewise, they were *relevant* and thereby enabled alignment of short-term and midterm action plans through concrete on-the-ground examples relevant to the Xinshe SEPLS. Finally, being comprehensible, they employed the "health check" and "SEPLS well-being" concepts as helpful terms to communicate the monitoring and evaluation processes to all multiple stakeholders.

The *challenges* that the Xinshe Initiative would need to address in the upcoming years include keeping up the momentum and ensuring the long-term engagement of all multiple stakeholders, staying flexible in the face of uncertainties (such as COVID-19), and communicating the values of the Xinshe Initiative to newly returned migrant youth and outsiders (e.g. avoiding the use of pesticides and herbicides, adopting eco-friendly waste management habits, learning to work with the government agencies).

The *opportunities* largely stem from the solution-oriented, positive-thinking approach of the local people and other stakeholders. Ways forward include conducting follow-up RAWs (in 2023) to observe the emergence of new synergy patterns and to readjust the action tasks; engagement of new stakeholders from outside of the COA system (e.g. Taiwan Fisheries Research Institute, Hualien County Government, Hualien Tourism Department, Fengbin Township Office) to address the newly emerged risks and resources; and building upon the social capital within and the partnership between the Fuxing and Xinshe communities (e.g. fostering the Jialang River dialogue).

Acknowledgements This case study would not have been possible without the kind support and help of many individuals and organisations. We would like to express our appreciation to the Council of Agriculture Executive Yuan who provided the funding to carry out this research. Our thanks and gratitude also go to the local people of Xinshe Village, Hualien County, who kindly welcomed and assisted us throughout our fieldwork.

Appendix A

A set of 20 localised indicators of resilience in Xinshe SEPLS *(translated from the Chinese language)*

Five perspectives of the Satoyama Initiative	Localised indicators of resilience in Xinshe SEPLS[a]
A. Ecosystem health and connectivity	A1. Diversity and multifunctionality of Xinshe SEPLS elements
	A2. Connectivity between Xinshe SEPLS elements
	A3. Xinshe SEPLS recovery and regeneration from natural hazards
	A4. Xinshe SEPLS protection and restoration from socio-economic threats
B. Sustainable resource use	B1. Diversity and consumption of locally grown food
	B2. Conservation and breeding of local crops and native species
	B3. Environmentally friendly farming practices with a circular resource use
	B4. Sustainable use of common resources (forest, river, and ocean)

(continued)

Five perspectives of the Satoyama Initiative	Localised indicators of resilience in Xinshe SEPLS[a]
C. Traditions and innovation	C1. Transfer of traditional ecological knowledge (TEK)
	C2. Documentation of traditional ecological knowledge (TEK)
	C3. Innovations and appropriate technology for agriculture and conservation
	C4. Recognition and respect for traditional ecological knowledge (TEK) of men and women
D. Multi-stakeholder governance	D1. Rights in relation to the management of Xinshe SEPLS natural resources
	D2. Multi-stakeholder governance mechanism in Xinshe SEPLS
	D3. Social capital of Fuxing/Xinshe communities
	D4. Community participation and benefit sharing in Fuxing/Xinshe communities
E. Sustainable livelihoods	E1. Local infrastructure
	E2. Human health and environmental conditions
	E3. Xinshe SEPLS ecosystem-based income diversity and marketing ability
	E4. Community spatial and temporal resource-use adaptive capacity

[a]*Comprehensive*: the "explanation-evaluation-discussion" model of conducting RAWs encouraged the participants to think of both risks and resources within each of the five perspectives; *Relevant*: local examples from the short-term action plan and the locals' insights from semi-structured interviews were added under each Localised Indicator; *Comprehensible*: pictorial explanation and clarifying question(s) were added under each indicator (see Appendix C for more details)

Appendix B

Three stages of the 2020 Resilience Assessment Workshops (RAWs) process

RAWs stage	Time	Description of activities
Pre-RAWs	April–June 2020	• Analysis of short-term action plan (2016–2019) of Xinshe Eco-Agriculture Initiative against the original set of 20 Indicators of Resilience in SEPLS (Bergamini et al., 2014) and development of first draft of 20 Localised Indicators of Resilience (the Localised Indicators) in Xinshe SEPLS (Sun et al., 2020) • Preparatory workshop with regional government agency members of multi-stakeholder platform (HDARES, HFDOFB, HBSWC, and EBAFA): matching action tasks of short-term action plan against first draft, collective development of second draft of localised indicators • 14 one-on-one semi-structured interviews with local residents from Fuxing and Xinshe tribes: test running and editing of second draft on "comprehensive-relevant-comprehensible" basis

(continued)

RAWs stage	Time	Description of activities
		• Completion of third (final) draft of localised indicators based on the analysis of previous steps
Result: A finalised set of 20 localised indicators of resilience in the Xinshe SEPLS		
Core RAWs	June–September 2020	• 5 + 1 community-based consecutive RAWs in Fuxing and Xinshe communities (12 RAWs in total): five (5) workshops in each community were RAWs based on the five perspectives (one for each perspective), one (1) was a summary workshop in each community for discussion of assessment results • Set of 20 localised indicators and perspective-related evaluation sheets (five in total) used during all five RAWs (see Appendix C) • Same group of participants from each tribe completed RAWs: 12 community members from Fuxing tribe and 8 from Xinshe tribe • Procedure of each RAW based on "explanation-scoring-discussion" model (Lee et al., 2020) with a particular emphasis on risks and resources within each perspective • Follow-up analysis carried out upon completion of each RAW and reported at the beginning of the following RAW in order to collect additional suggestions and opinions from community members
Result: A community-based perception of risks and resources in the Xinshe SEPLS		
Post-RAWs	September–December 2020	• Drafting of first version of midterm action plan (2021–2026) for the Xinshe initiative based on RAWs' results • Preparatory workshop with regional government agency members of multi-stakeholder platform (HDARES, HFDOFB, HBSWC, and EBAFA): communication of results of RAWs, division of responsibilities for action tasks in first draft of midterm action plan • Joint workshop open to all community members from Fuxing and Xinshe tribes with invited heads of the regional government agency members of the multi-stakeholder platform (HDARES, HFDOFB, HBSWC, and EBAFA), Xinshe Primary School (local school), and Fengbin Township Office (local government): communicating results of RAWs with a special emphasis on risks and resources of Xinshe SEPLS, collective development of second draft of midterm action plan via a Q&A session between local community members and government officials • Two task force meetings and a semi-annual multi-stakeholder platform meeting: adding final edits to the division of responsibilities for action tasks and adoption of third (final) version of midterm action plan
Result: A bottom-up midterm action plan (2021–2026) for the Xinshe "Forest-River-Village-Ocean" Eco-Agriculture Initiative		

Appendix C

Sample of the Resilience Assessment Workshops (RAWs) Evaluation Sheet, Perspective A *(Translated from the Chinese Language)* (Source: Prepared by Authors)
    Participant's name:_____

Localised Indicators	A1. Diversity and multifunctionality of the Xinshe SEPLS elements	A2. Connectivity between the Xinshe SEPLS elements	A3. The Xinshe SEPLS recovery and regeneration from natural hazards	A4. The Xinshe SEPLS protection and restoration from man-made (socio-economic) threats
Explanatory pictures				
Evaluation questions	Does the Xinshe SEPLS contain diverse natural and man-made ecosystem elements? Are these elements healthy and well functioning?	Are the forest, river, farmlands, and ocean well connected? Are there any constraints to the Xinshe SEPLS connectivity?	Does the Xinshe SEPLS recover and regenerate quickly from natural hazards?	Does the Xinshe SEPLS possess effective protection and restoration measures from man-made (socio-economic) threats?
Scores	☆☆☆☆☆	☆☆☆☆☆	☆☆☆☆☆	☆☆☆☆☆
Relevant action tasks of the Xinshe initiative	A1. Investigate and enhance diversity and multifunctionality of the Xinshe SEPLS elements	A2. Investigate and enhance connectivity between the Xinshe SEPLS elements	A3. Assess the susceptibility of the Xinshe SEPLS to natural hazards and enhance its recovery and regeneration capacity	A4. Protect and restore the Xinshe SEPLS from man-made (socio-economic) threats
*Indicator-relevant local examples of specific action tasks (**please score in order of priority**)*	☐ Conduct regular biodiversity checks and share results with the local people ☐ Improve protection of water quality and quantity of the Jialang River ☐ Enhance man-made landscape diversity	☐ Investigate and communicate the role of eco-agriculture for the SEPLS health and forest-river-village-ocean connectivity (e.g. via coral reef checks and wetland buffer zone construction)	☐ Improve monitoring and disaster response to coastal erosion (including eco-friendly engineering projects) ☐ Strengthen forest and stream (the Jialang River) patrol to monitor the risks	☐ Develop community-based river protection mechanism to prevent non-locals from fishing in the Jialang River ☐ Prevent "construction invasions" that negatively impact local

(continued)

Localised Indicators	A1. Diversity and multifunctionality of the Xinshe SEPLS elements	A2. Connectivity between the Xinshe SEPLS elements	A3. The Xinshe SEPLS recovery and regeneration from natural hazards	A4. The Xinshe SEPLS protection and restoration from man-made (socio-economic) threats
	(e.g. home gardens, rice paddies, green buffers, hedgerow plants) □ Learn about the Xinshe SEPLS values from tribal history and culture (e.g. farming, fishing, hunting, gathering, arts and crafts) and apply them to everyday life	□ Improve protection of water quality and quantity of the Jialang River □ Carry out eco-friendly restoration projects (e.g. terraced irrigation, drainage systems and ponds, repairing terraces, dams, and agricultural roads) □ Conduct reforestation and restoration of degraded forest land □ Develop measures to recover the spawning migration of fish and shrimp	of erosion, landslides, and floods □ Set up a "crisis team" in response to typhoons □ Remove weeds that obstruct the water flow of the Jialang River (flood prevention) □ Develop affordable and maintainable methods to deal with crop-raiding species (e.g. Formosan macaques, wild boar, wild hare, and barking deer) □ Prevent and control invasive animal and plant species (e.g. *Mikania micrantha*)	ecology/scenery (e.g. utility poles, high-voltage street-lights, and solar panels) □ Educate the local farmers to abstain from the use of pesticides and herbicides □ Strengthen tourist environmental education to prevent destructive behaviour (e.g. littering, trampling, use of DEET mosquito sprays, and SPF lotions) □ Strengthen environmental protection awareness of the local residents (e.g. household waste sorting, prohibiting burning of garbage, reducing single-use plastic) □ Prevent destructive fishing methods (e.g. trawling, drift nets) □ Strengthen forest patrol to prevent illegal logging

Appendix D

The midterm action plan for the Xinshe "Forest-River-Village-Ocean" Eco-Agriculture Initiative *(translated from the Chinese language)* (source: prepared by authors)

Division of responsibilities between the multiple stakeholders (★ leading role; ✓ supporting role)

Five Action Perspectives	10 Thematic priority areas (2021-2026)	Checklist of 44 specific action tasks (2021-2023) *(to be adjusted at the end of each year, to be re-assessed by the follow-up RAWs in 2023)*	Amis Fuxing Tribe	Kavalan Xinshe Tribe	Xinshe Primary School	Fengbin Township Office	Hualien County Government	HDARES	HFDOFB	HBSWC	EBAFA	Irrigation Agency	NDHU	Hualien Tourism Department	Other
A: Protect health and connectivity of the Xinshe SEPLS	"River": Protect the Jialang River resources and set up a community-based river protection mechanism	Maintain drinking and irrigation water quality (including sewage discharge management) and quantity (including adequate supply of drinking water for each household), protect the Jialang River's water source	★	★		★						★			
		Improve the spawning migration passage for fish and shrimp ("river-sea" connectivity)	✓	✓			★		✓						
		Sustainably use local fish and shrimp resources, set up a community-based river protection mechanism and erect notice boards	★	★		★	✓		★						
		Organise de-weeding of the riverbed when needed based on consensus between the local residents (preferably no dredging)	★	★			★			★					
		Strengthen local residents' knowledge about river protection laws, regulations and competent authorities	✓	✓		✓	★		✓						
	"Ocean": Protect the ocean resources and set up a community-based ocean protection mechanism	Monitor coastal erosion, consider eco-friendly coastal engineering projects	✓	★		✓	★						✓		
		Prevent destructive fishing or collection methods	★	★			★								
		Strengthen community-based sustainable use of marine resources in traditional territories	★	★			★								
		Monitor the health of marine ecosystem (including coral reef and fish checks)	★	★			✓								
		Strengthen local residents' knowledge about ocean protection laws, regulations and competent authorities	★	★		★	★						✓		
B: Sustainably use natural resources of the	"Eco-agricultural production": promote organic, environmental[ly friendly]	Promote expansion of organically-farmed zones, assist farmers with installation of small processing units (upon request)	★	★				✓	✓		★				
		Provide extension and training services on organic, environmentally-friendly and circular farming techniques	★	★				★	✓		✓				

Xinshe SEPLS	...lly-friendly and circular agriculture	A ➤ Discourage the use of pesticides and herbicides, set up a de-weeding team (funded on a cash rebate basis)
		A ➤ Prevent and control crop-raiding species (e.g. Formosan macaques, wild boar, wild hare, and barking deer)
		A ➤ Prevent "construction invasions" (streetlights, solar panels) from harming crops
		A ➤ Provide extension and training services on edible wild plants and agri-food education
		A ➤ Sustainably use and consume local agro-forestry products
	"Native species"; preserve and reintroduce native plant and animal varieties	A ➤ Combine TEK and modern science to preserve and reintroduce local native species of plants and animals
		A ➤ Study and promote crops that are ecologically friendly and suitable for local terrain, climate, and hydrology
		A ➤ Establish tribal seed banks and conservation gardens
C: Weaving TEK and modern science	"Traditional knowledge"; transfer and document TEK and culture	A ➤ Transfer TEK and practical skills of the local elderly (men and women)
		A ➤ Educate local youths with the help of Xinshe Primary School's place-based curriculum
		A ➤ Encourage local residents and experts to document TEK (based on the principle of maintaining tribal intellectual property)
		A ➤ Establish local TEK Library and set up a digital data storage platform to make documented TEK accessible to all local users
	"Return of migrant youth"; create supporting conditions for adaptation	A ➤ Conduct vocational training on Indigenous languages (Amis, Kavalan and Bunun) and tribal culture
		A ➤ Create local job opportunities based on specific abilities of the returned migrant youth (including relevant capacity-building activities)
		A ➤ Take advantage of the good reputation of Xinshe Primary School to entice the migrant youth to return with their families
D: Enhance community-based collaborative governance	"Social capital"; enhance participation, leadership and benefit sharing within the tribes	A ➤ Establish digital database/inventory of various skills, talents, experiences, and knowledge of tribal members (including both current residents and out-migrated youth)
		A ➤ Integrate various groups and organisations within the tribes to collectively discuss and solve problems
		A ➤ Strengthen communication and partnership with out-migrated youth and other tribal members residing elsewhere

Category / Strategy	Activity											
"Multi-stakeholder platform": enhance partnership among all SEPLS actors	➤ Strengthen local residents' understanding and recognition of the operational mechanism and the mid-term action plan of the Xinshe Initiative	★	★	★	✓	✓	✓	✓	✓		✓	
	➤ Develop an interactive and diversified operation of the multi-stakeholder platform for the Xinshe Initiative	★	★	✓	✓	✓	✓	✓	✓		✓	
	➤ Strengthen communication, problem solving and partnership for common objectives between Fuxing and Xinshe Tribes	★	✓		✓	✓	✓					
	➤ Strengthen communication, problem solving and partnership for common objectives between the tribes and the government agencies	★	★	✓	✓	✓	✓	✓	✓	✓		
	➤ Encourage proactive involvement of Fengbin Township Office	✓	✓	★	✓	✓	✓	✓	✓		✓	
"Marketing": promote marketing of local eco-agricultural produce	➤ Develop value-added products based on their unique *satoyama* characteristics (including agriculture, crafts, culinary art, tourism, etc.) and improve access to various marketing channels	★	★	✓	✓	✓	✓	★				★
	➤ Assist the tribes with the acquisition of processing equipment and learning about processing technologies	★	★	✓	★			★				
	➤ Assist the tribes with registration for relevant organic and environmentally-friendly certification schemes/labels	★	★	✓	★			★				
	➤ Provide extension and training services to enhance the marketing skills of the local residents	★	★	✓	✓	✓	✓	★				
E: Enhance local livelihoods and well-being / **"Safety": strengthen disaster prevention measures, improve safety and elderly-friendly infrastructure**	➤ Improve road safety (including road lighting, installation of grille gutter covers, speed limit signs, and traffic lights) and convenience (including potential road diversion to avoid local cemetery)	✓	✓	✓			★					
	➤ Improve transportation services and day care for the elderly	★	★	★	✓	★						
	➤ Improve elderly-friendly emergency shelter infrastructure	✓	✓	★	✓							
	➤ Strengthen environmental education for tourists to reduce damage to the Xinshe SEPLS environment	★	★	✓	✓							★
	➤ Conduct disaster risk assessment, monitoring and prevention (including forest, stream, coastal, and post-typhoon patrol)	★	★	★	✓		★					

HDARES Hualien District Agricultural Research and Extension Station, *HFDOFB* Hualien Forest District Office, *HBSWC* Hualien Branch of Soil and Water Conservation Bureau, *EBAFA* Eastern Region Branch of Agriculture and Food Agency, *NDHU* National Dong Hwa University, *RAWs* resilience assessment workshops, *SEPLS* socio-ecological production landscape and seascape, *TEK* traditional ecological knowledge

References

Bergamini, N., Dunbar, W., Eyzaguirre, P., Ichikawa, K., Matsumoto, I., Mijatovic, D., Morimoto, Y., Remple, N., Salvemini, D., Suzuki, W. & Vernooy, R. (2014) *Toolkit for the indicators of resilience in socio-ecological production landscapes and seascapes.* UNU-IAS, Biodiversity International, IGES & UNDP.

Brown, K. (2016). *Resilience, development, and global change.* Routledge.

Carpenter, S., Walker, B., Anderies, J. M., & Abel, N. (2001). From metaphor to measurement: Resilience of what to what? *Ecosystems, 4*(8), 765–781.

Convention on Biological Diversity (CBD). (2020). *Zero draft of the post-2020 global biodiversity framework.*

Dodge, R., Daly, A., Huyton, J., & Sanders, L. (2012). The challenge of defining well-being. *International Journal of Wellbeing, 2*(3), 222–235.

Ford, J. D., King, N., Galappaththi, E. K., Pearce, T., McDowell, G., & Harper, S. L. (2020). The resilience of indigenous peoples to environmental change. *One Earth, 2*(6), 532–543.

Forgeard, M. J., Jayawickreme, E., Kern, M., & Seligman, M. E. P. (2011). Doing the right thing: measuring well-being for public policy. *International Journal of Wellbeing, 1*(1), 79–106.

Freepik (2021) Free vectors graphic resources [Human body systems] viewed 31 October 2021. Retrieved from https://www.freepik.com/vectors/human.

Google. (2021) Google Maps [Xinshe Village, Fengbin Township, Hualien County, Chinese Taipei] viewed 31 October 2021. Retrieved from https://goo.gl/maps/rpCBn1w98Ni69Egm9.

IPSI Secretariat. (2014). *The International Partnership for the Satoyama Initiative (IPSI): Working towards societies in harmony with nature.* United Nations University Institute for the Advanced Study of Sustainability.

Karimova, P. G. (2021). An integrated landscape approach for revitalisation of indigenous socio-ecological production landscape and seascape in Xinshe Village, Hualien County, Taiwan. *PANORAMA: Solutions for a healthy planet* online platform, viewed 1 August 2021. Retrieved from https://panorama.solutions/en/solution/integrated-landscape-approach-revitalisation-indigenous-socio-ecological-production.

Karimova, P. G., & Lee, K. C. 2021, The good, the bad and the adaptive: Resilient local solutions to tourism related system shifts in Eastern rural Taiwan. *Taiwan Insight* online magazine, viewed 27 August 2021. Retrieved from https://taiwaninsight.org/2021/07/31/the-good-the-bad-and-the-adaptive-resilient-local-solutions-to-tourism-related-system-shifts-in-eastern-rural-taiwan/.

Lee, K. C. (2016). The Satoyama initiative and eco-agriculture at a landscape scale. *Landscape Conservation Newsletter, 42,* 12–18. (in Chinese).

Lee, K. C., Karimova, P. G., & Yan, S. Y. (2019). Towards an integrated multi-stakeholder landscape approach to reconciling values and enhancing synergies: A case study in Taiwan. In UNU-IAS & IGES (Ed.), *Understanding the multiple values associated with sustainable use in socio-ecological production landscapes and seascapes (SEPLS)* (Satoyama initiative thematic review) (Vol. 5, pp. 118–133). United Nations University Institute for the Advanced Study of Sustainability.

Lee, K. C., Karimova, P. G., Yan, S. Y., & Li, Y. S. (2020). Resilience assessment workshops: An instrument for enhancing community-based conservation and monitoring of rural landscapes. *Sustainability, 12*(1), 408–422.

Nishi, M., & Yamazaki, M. (2020). Landscape approaches for the post-2020 biodiversity agenda: Perspectives from socio-ecological production landscapes and seascapes. UNU-IAS policy brief series. United Nations University.

Oxford English Dictionary. (2021). 'Well-being' definition, viewed 5 October 2021. Retrieved from https://www.oed.com/oed2/00282689;jsessionid=6C04BC77AD21CC7D7034E30 96CDC0AD4.

Scherr, S., & McNeely, J. (2008). Biodiversity conservation and agricultural sustainability: Towards a new paradigm of 'eco-agriculture' landscapes. *Philosophical Transactions of the Royal Society, 363*, 477–494.

Sun, X. T., Yan, S. Y., & Lee, K. C. (2020). Localised indicators of resilience for adaptive management: Building up resilient SEPLS in Xinshe Village, Hualien County. *Taiwan Forestry Journal, 46*(6), 58–80. (in Chinese).

The opinions expressed in this chapter are those of the author(s) and do not necessarily reflect the views of UNU-IAS, its Board of Directors, or the countries they represent.

Open Access This chapter is licenced under the terms of the Creative Commons Attribution 3.0 IGO Licence (http://creativecommons.org/licenses/by/3.0/igo/), which permits use, sharing, adaptation, distribution and reproduction in any medium or format, as long as you give appropriate credit to UNU-IAS, provide a link to the Creative Commons licence and indicate if changes were made.

The use of the UNU-IAS name and logo, shall be subject to a separate written licence agreement between UNU-IAS and the user and is not authorised as part of this CC BY 3.0 IGO licence. Note that the link provided above includes additional terms and conditions of the licence.

The images or other third party material in this chapter are included in the chapter's Creative Commons licence, unless indicated otherwise in a credit line to the material. If material is not included in the chapter's Creative Commons licence and your intended use is not permitted by statutory regulation or exceeds the permitted use, you will need to obtain permission directly from the copyright holder.

Chapter 5

"To Take Care of the Land Means Taking Care of Ourselves": Local Perceptions on Human and Environmental Health in a High Agro-Biodiversity Landscape in the Yucatan Peninsula

María Elena Méndez-López, María Fernanda Cepeda-González, Karla Juliana Rodríguez-Robayo, Lilian Juárez-Téllez, Mariana Rivera-De Velasco, Rosa Martha Peralta-Blanco, Nicolás Chan-Chuc, Andrea A. Serrano-Ysunza, R. Antonio Riveros-Cañas, Oscar G. Sánchez-Siordia, and Sebastien Proust

Abstract The Forest and Milpa Landscape (FML) is a territory comprising 64 municipalities in the Yucatan Peninsula where the rainforest and the *milpa* system coexist. The ecosystems that predominate in the FML are sub-deciduous and subtropical evergreen forests, which represent an essential carbon reservoir worldwide. The use of natural resources for food security of FML families is associated with the *milpa*, which is a system that depends on the rainfall and the soil's ability to retain water. Within the framework of the 2020–2030 Country Strategy of the GEF

M. E. Méndez-López (✉) · L. Juárez-Téllez
CONACYT—Centro de Investigación en Ciencias de Información Geoespacial, Mérida, Yucatán, Mexico
e-mail: emendez@centrogeo.edu.mx

M. F. Cepeda-González
Capacitación y Asesorías Ambientales, Mérida, Yucatán, Mexico

K. J. Rodríguez-Robayo
CONACYT—Centro de Investigación en Ciencias de Información Geoespacial, Mérida, Yucatán, Mexico

Centro de Investigación Tibaitatá AGROSAVIA, Mosquera, Colombia

M. Rivera-De Velasco · R. M. Peralta-Blanco · N. Chan-Chuc · O. G. Sánchez-Siordia
Centro de Investigación en Ciencias de Información Geoespacial, Mérida, Yucatán, Mexico

A. A. Serrano-Ysunza · S. Proust
The GEF Small Grants Programme, United Nations Development Programme, Mérida, Yucatán, Mexico

R. A. Riveros-Cañas
Independent Consultant, Mérida, Yucatán, Mexico

© The Author(s) 2022
M. Nishi et al. (eds.), *Biodiversity-Health-Sustainability Nexus in Socio-Ecological Production Landscapes and Seascapes (SEPLS)*, Satoyama Initiative Thematic Review, https://doi.org/10.1007/978-981-16-9893-4_5

Small Grants Programme (SGP), 20 indicators associated with the FML's resilience were evaluated through a participatory approach. The methodological route consisted of adapting the Toolkit for the Indicators of Resilience in Socio-ecological Production Landscapes and Seascapes (SEPLS). A topic that generated much concern among participants was human health. The reflection generated around this indicator recognised problems associated with water contamination by agrochemicals and changes in diet, resulting in recurrent diseases, such as diabetes, hypertension, and obesity. The solutions proposed by the small producers are linked to the sustainable management of ecosystems and education on values towards traditional and agroecological food production.

Keywords Forest and *milpa* landscape · Resilience · SEPLS · *Milpa* · Yucatan Peninsula

1 Introduction

The Forest and Milpa Landscape (FML) is a territory comprising 64 municipalities distributed across the three states of the Yucatan Peninsula (YP) in Mexico: Campeche, Quintana Roo, and Yucatan. The FML is a region where the rainforest and the *milpa* system coexist (Fig. 5.1 and Table 5.1), and is one of the five landscapes where the GEF Small Grants Programme (SGP) operates in Mexico. The names given to each landscape aim to describe the ecosystems and traditional production activities of the region, creating an identity among communities. The FML is part of the participatory creation of the 2020–2030 Country Strategy of the SGP.

The SGP adopted a community-based landscape approach during its sixth operational phase (OP6), which recognises that community-based organisations are the driving force in rural development strategies and must take the lead in project planning, landscape governance, project execution, and monitoring. This approach is part of a strategic initiative to promote conservation and sustainable use of natural resources and ecosystems.

About one million people inhabit the FML, of which 48% is considered an economically active population (Instituto Nacional de Estadística y Geografía (INEGI), 2015a). More than three-quarters of the municipalities have upwards of 10% of residents engaged in subsistence agriculture. On average, 29% of the economically active and occupied population is engaged in natural resource-use activities (Instituto Nacional de Estadística y Geografía (INEGI), 2015a). In the FML, 45% of the population speaks an indigenous language, mainly Maya (Instituto Nacional de Pueblos Indígenas, 2015). The agro-biodiversity of *milpas* is an essential element for food security of indigenous families in the FML (Salazar Barrientos & Magaña Magaña, 2016).

The *milpa* is an agroecosystem based on rotational cultivation under the slash-and-burn technique, which depends on the seasonal rainfall and the soil's ability to retain water (Martínez et al., 2017; Salazar Barrientos et al., 2016). It makes up a matrix of polycultures, family gardens, and fragments of natural vegetation

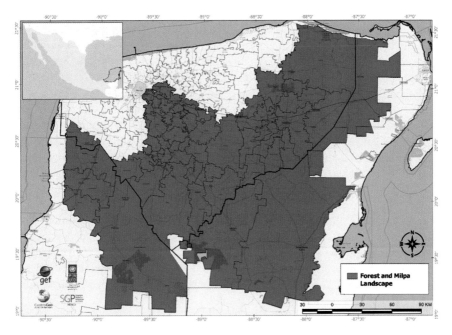

Fig. 5.1 Regionalisation of the forest and milpa landscape (source: prepared by Rosa Martha Peralta, Instituto Nacional de Estadística y Geografía (INEGI), 2015b; Instituto Nacional de Estadística y Geografía (INEGI), 2016; Comisión Nacional para el Conocimiento y Uso de la Biodiversidad (CONABIO), 2015)

Table 5.1 Basic information of the study area

Country	Mexico
Province	Campeche, Quintana Roo, and Yucatan
District	n.a.
Size of geographical area (hectare)	5,268,000
Number of direct beneficiaries (persons)	n.a.
Number of indirect beneficiaries (persons)[2]	139,200
Dominant ethnicity(ies), if appropriate	Maya
Size of the case study/project area (hectare)	n.a.
Geographic coordinates (latitude, longitude)	20°12′18.0″N, 88°43′48.0″W

(Zizumbo-Villarreal & Colunga-García, 2017; Mariaca, 2015). The agro-biodiversity of the *milpa* is based on the so-called three sisters, corn, beans, and squash (Odum & Sarmiento, 1998), but comprises more than 30 species of other edible and medicinal plants (Toledo et al., 2003; Salazar Barrientos et al., 2016) along with other forbs and grasses known in Spanish as *arvenses*. The *milpa* in the FML has led to increased landscape diversity due to multistage and successional pathways of native secondary growth vegetation (Terán, 2010).

The ecosystems that predominate in the FML are sub-deciduous and subtropical evergreen forests, which represent an essential carbon sink worldwide. Yet, they are in a highly vulnerable status mainly due to the increase in mechanised agriculture and cattle ranching caused by the high demand for food (Aide et al., 2013; Comisión Nacional de Áreas Naturales Protegidas, 2020).

Regarding water, the FML has a peculiar characteristic: surface streams are almost inexistent. Water runs underground, and openings to the exterior are sink-holes or dolines called *cenotes*. The terrestrial surface consists of porous limestone rock with high permeability and transmissibility that allows rain to pass easily to the underground; in addition, rock fractures facilitate water flow, and other liquids poured onto their surfaces flow through the fissures. Thus, the FML aquifer is highly vulnerable to pollutants (Hernández & Ortega, 2017).

The FML is a region with shallow stony soils, rainfall patterns with 6 months of low rain, and a high incidence of hurricanes. Nevertheless, Mayan communities in the Yucatan Peninsula have adapted the *milpa* system to these adverse conditions (Toledo et al., 2008). This socio-environmental resilience is associated with the Yucatan's Mayan communities' multi-use strategies of nature that allow for the use of a variety of natural resources, at domestic units, both for subsistence purposes and for local and regional economic exchanges (Barrera-Bassols & Toledo, 2005; García-Frapolli et al., 2008; Toledo & Barrera-Bassols, 2011; Pinto & Barrios, 2015).

Today, the FML faces various threats that have diminished its resilience capacity. In the cultural sphere, a pattern of migration of young people to the Mayan Riviera has blocked the heritage of local knowledge associated with the *milpa* and their sustainable management (Rodríguez-Robayo et al., 2020). On the other hand, public policies have been implemented to make the *milpa* more profitable by promoting technology packages with genetically enhanced seeds and agrochemicals (Gutiérrez Núñez, 2020). Most of these chemicals are organochlorinated pesticides (OCPs) that mimic the functions of natural human hormones once they enter the body through water taken from contaminated wells and *cenotes*, and *milpa* foods consumed, by skin exposure or by inhalation (Polanco, Araujo, et al., 2018a). Pesticide-contaminated water in the FML is also linked to breast and uterine cancer and the presence of organochlorine substances in the breast milk of Mayan women (Polanco et al., 2017; Polanco, López, et al., 2018b).

The FML, as we have seen, is a landscape with multiple vulnerabilities. In this chapter, we want to share the vision of the FML inhabitants—their concerns over problems related to human health that they are facing today, and the strategies with which they aspire to overcome them.

We share reflections on the resilience of the FML from the perspective of the community members that inhabit it, namely, the peasants. A deliberation exercise was carried out by adapting the Toolkit for the Indicators of Resilience in Socio-ecological Production Landscapes and Seascapes (SEPLS) (Bergamini et al., 2014). These indicators seek to enhance local communities' sense of proprietorship over landscape management processes and encourage them to think about how the landscape's resilience can be improved. While indicators address different aspects of resilience, this study focuses on the issue of human and environmental health.

2　Methods

Within the framework of the 2020–2030 Country Strategy of the Small Grants Programme (SGP), two workshops were held in 2019, attended by a total of 40 peasants (18 women/22 men) living in the FML. The objective of the Country Strategy is to conduct a collective reflection on the status of production landscapes in Mexico, and propose plans based on local knowledge to improve those aspects where problems exist.

It was first necessary to identify stakeholders from the three states of the Yucatan Peninsula to invite to participate. We made a list of community leaders who have developed conservation and sustainability projects within the FML, and then we contacted these leaders and asked them to extend invitations to those interested in attending. Young people, women, and men were invited to these workshops to ensure that the exercise represented the various generations and genders in society.

The objective of the workshops was to evaluate 20 indicators that were adapted from the SEPLS Toolkit (Bergamini et al., 2014). This toolkit was used to generate resilience indicators from the perspective of the peasants who use and inhabit this socio-ecological production landscape.

Because participants came from different FML regions, sub-landscapes were defined using a participatory mapping technique, where producers themselves determined the extent of the territory on which they could make the assessment. Six sub-landscapes of the FML were evaluated utilising the indicators (Fig. 5.2).

Each sub-landscape was represented by a team of between five and six participants, who evaluated the set of indicators. The help of three Mayan-Spanish translators was engaged to ensure an inclusive process.

The Metaplan technique was used to evaluate the indicators. The process consisted of developing key questions to achieve a brainstorming process, followed by collective reflection and, finally, reaching a consensus on the score of each indicator using the Likert scale, with scores ranging from 1 to 5, where 1 is very low and 5 is very high. To ensure an accurate numerical value for each indicator, each participant was given a template with the five ranges (Fig. 5.3), and participants were asked to vote for the ranking category in which they considered each indicator to be, based on the discussion in the previous collective reflection. The values were averaged within the team and then between groups to get each indicator's overall value. Participants assessed five groups of indicators: (a) *heterogeneity and landscape protection*, (b) *agrodiversity and shared natural resources*, (c) *traditional knowledge and innovation*, (d) *governance and social equity*, and (e) *livelihoods*.

Each workshop lasted 2 days. During the first day, each team devoted an hour and a half to each group of indicators, with recess and recreation spaces to avoid fatigue.

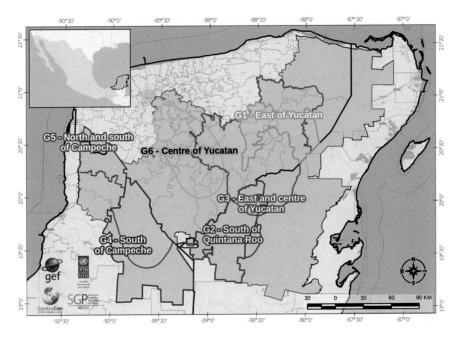

Fig. 5.2 Sub-landscapes of the FML as defined by six groups of participants from different regions (prepared by Rosa Martha Peralta)

Fig. 5.3 Ranges used to evaluate indicators of resilience (source: prepared by authors)

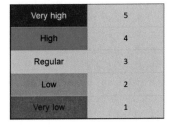

Very high	5
High	4
Regular	3
Low	2
Very low	1

On the second day, the results were discussed for each sub-landscape, and some strategies were developed to address indicators that obtained low ratings.

3 Results

3.1 Indicators of Resilience

The results are shown in Table 5.2. As noted, only 4 out of 20 indicators received a "high" rating, with most of them (13) evaluated as "regular," and only 3 put into the "low" category. In the first group, *landscape heterogeneity and ecosystem*

Table 5.2 Results of indicators of resilience (adapted from SEPLS Toolkit) for FML (source: prepared by authors)

Indicator	Evaluation
Heterogeneity and landscape protection	
1) Heterogeneity of the terrestrial/marine landscape	High
2) Ecosystem protection	Regular
3) Ecological interactions between different components of the marine/terrestrial landscape	Regular
4) Recovery and regeneration of the marine/terrestrial landscape	Regular
Agrodiversity and shared natural resources	
5) Diversity of local production systems	Regular
6) Maintenance and use of native varieties and species	Regular
7) Sustainable management of shared resources	Regular
Traditional knowledge and innovation	
8) Innovation in production practices and conservation	Low
9) Traditional biodiversity-related knowledge	Regular
10) Systematisation of biodiversity-associated knowledge	Regular
11) Women's knowledge	High
Governance and social equity	
12) Land/water rights and the management of other natural resources	Regular
13) Community governance of the marine/terrestrial landscape	Low
14) Synergy of social capital in the marine/terrestrial landscape	Regular
15) Social equity (includes gender equity)	Regular
Livelihoods	
16) Socioeconomic infrastructure	Regular
17) Human health and environmental conditions	Low
18) Productive diversification	High
19) Biodiversity-based livelihoods	High
20) Socio-ecological mobility	Regular

protection, one indicator was evaluated as high, and three were regular. In the second group, *agrodiversity and shared natural resources*, all three indicators were deemed to be regular. In the group of *traditional knowledge and innovation* indicators, one indicator was low, two regular, and one high. Three indicators were regular and one low in *governance and equity*. Finally, in the *livelihoods* group, two indicators were scored high, two regular, and one low (Table 5.2).

To address the issue of human and environmental health, we delved into the indicator that obtained the lowest value in both workshops: human health and environmental conditions (livelihoods group). Reasons for this focus include this indicator being an issue to which participants gave much weight in collective reflections, and also one that is associated with other indicators whose values were regular or low: socioeconomic infrastructure, traditional knowledge, community governance, and local production diversity.

The questions that were asked to participants in reflecting on human health and environmental conditions were the following: (a) Are our communities healthy? (b) Is there any link between the diseases we have in the community and the environmental conditions (soil, *cenotes*, forests)?

For the first question, answers were negative for all teams. The most mentioned diseases in the FML communities were stomach infections, diabetes, cancer, kidney problems, and hypertension. Other isolated mentions were infertility, anaemia, and headaches, among others (Table 5.3).

Answers to the second question were all positive regarding the relationship between disease and environmental health. Participants linked diseases such as stomach infections and cancer to water and air pollution caused by the lack of landfills and consequent burning of litter. Causes of diarrhoea mentioned included the increasing hot weather in recent years and water contamination due to the establishment of pig farms. Likewise, cancer was linked to water contamination by toxic residues from mechanised agriculture and excess of hormones in food.

Participants explained that diabetes is related to switching from a *milpa*-based diet (corn, bean, squash, bush meat, and honey, among other products) to an industrialised diet. Participants noted that the consumption of junk food and high-sugar carbonated beverages has increased in the last years in their communities. Participants associate this phenomenon with the indicator for traditional knowledge related to biodiversity, concerning which reflection focused on young people in communities losing interest in the *milpa* and its rituals: "They see traditional knowledge as backwardness". This trend is strengthened by cultural uprooting due to migration, resulting in the *milpa* foodscape and other local production systems not being valued and a rising preference for industrialised products from the global market.

Other environmental health issues associated with the aetiology of diseases included water in communities containing a lot of chlorine; high deforestation rate of remnant forest patches; increment of disease vectors due to increased heat; and planting of GMOs.

Table 5.3 Responses to reflection questions for *human health and environmental conditions* indicator, shown for each of the groups formed in the workshops

Name of group and sub-landscape (see Fig. 5.2)	Reflection questions	
	(a) Are our communities healthy? (b) What are the main diseases that you have? (When the number of mentions is >1, it is indicated in parentheses)	(c) Is there any link between the diseases in the community and the environmental conditions (soil, *cenotes*, forests)? (When the number of mentions is >1, it is indicated in the parenthesis)
Group 1. Venados amarillos **Sub-landscape:** East of Yucatan state	(a) No (b) – Diarrhoea – Fever – Cough – Stomach infections – Influenza – Pneumonia – Diabetes – Kidney stones – Rheumatism (2)	– Kidney stones are caused by drinking dirty water (2) – Water and air pollution (no municipal dump, the trash is burned and smells bad) – Food with chemicals gets us sick – The chemicals get into the hens (2) – Climate change causes coughing and flu – Climate is hotter causing fevers and diarrhoea
Group 2. Loros verdes **Sub-landscape:** South of Quintana Roo state	(a) No (b) – Diabetes (4) – High cholesterol – High triglycerides – Diarrhoea – Anaemia – Fever – Cancer (3) – Depression	– Bad food [a long time ago meals came from the *milpa*] (4) – Diseases from the consumption of junk food – Sudden changes in temperature [flu] (2) – Chlorinated water – Pig farms produce water pollution – High temperatures – Food with toxic fumigation residues (2)
Group 3: Jabalíes verdes **Sub-landscape:** East and centre of Yucatan state	(a) No (b) – Fever (3) – Diabetes – Cough – Headache – Cancer	– Plastic burning – The sun is stronger (2) – Junk meals – Agrochemicals – Because we eat foods that have chemicals – A lot of Coca-Cola consumption
Group 4: Xunan Kab **Sub-landscape:** South of Campeche state	(a) No (b) – Gastrointestinal problems (7) – Hypertension (4) – Increased infertility (2) – Diabetes (5)	– Food (vegetables and animals contaminated with chemicals and hormones (7) – Exposure to pesticides (3) – Deforestation (5) – Environmental/water/air pollution (5)

(continued)

Table 5.3 (continued)

Name of group and sub-landscape (see Fig. 5.2)	Reflection questions	
	(a) Are our communities healthy? (b) What are the main diseases that you have? (When the number of mentions is >1, it is indicated in parentheses)	(c) Is there any link between the diseases in the community and the environmental conditions (soil, *cenotes*, forests)? (When the number of mentions is >1, it is indicated in the parenthesis)
	– Skin poisoning (3) – Different types of cancer (4) – Increased children with learning disabilities – Kidney problems (2) – Increased fever cases (5) – Headaches – Increased cases of obesity – Early onset of puberty	– Factories (2) – Climate change (5)
Group 5: Che'el Azul **Sub-landscape:** North and south of Campeche state (coinciding in the south with group 4)	(a) No (b) – Diabetes – Diarrhoea (2) – Hypertension – Respiratory infections (2) – Salmonellosis – Kidney problems (2) – Embolism – Paralysis	– The chemicals used for agriculture – The climate is very unstable (2) – Outdoor garbage burning (2) – Chlorinated water (excess chlorine) – Street food consumption – Formerly there was a season of mosquitoes, now there are mosquitoes all the time
Group 6: Chaak naal téel **Sub-landscape:** Centre of Yucatan state	(a) No (b) – Diabetes (2) – Rheumatism – Cancer (2) – Anaemia – Chronic cough – Cirrhosis – Hypertension – Obesity	– Temperature changes (3) – Bad food (2) – Use of agrochemicals (2) – Lots of chlorine in the water – Environmental pollution by mega projects/farms – Transgenic planting

In addition to the health problems faced by FML communities, the socio-economic infrastructure indicator showed a lack of medical services (hospitals, ambulances, and access to medicines).

3.2 Local Strategies for Resilience

On the second day of the workshops, a reflection was made on the FML's priority aspects, based on the work done on the first day. As mentioned earlier, the human health indicator was chosen as a priority. Although many of the factors that cause health problems are outside the communities, proposals were made for strategies with the potential to address issues at the local level to increase the resilience of the FML.

Regarding the problem of diabetes, education is a crucial element; as one participant mentioned:

> Children do not know the food from the *milpas*, we have to work with them, tell them that corn, sweet potato and beans are the best food, and share with them what we know about the value of *milpa*.

The same person mentioned a project that he develops in schools to build organic gardens:

> Students learn the whole process and thus they value food because they know what it costs to get it.

The participant concludes by saying that the best thing people can do to avoid getting sick is "to take care of the land, that means taking care of ourselves, but it is something we have not understood".

A proposed strategy to prevent pesticide contamination in water and food was to implement innovations in production activities, as one participant suggested:

> We can take advantage of the excrements of animals from our yards to make fertiliser, and we also can use microorganisms; we can mix these techniques with the old practices of *milpa* to make agroecology.

The problem of pesticides from extensive agriculture is more complex, as it is caused by public policies that have promoted agrochemicals since the 1940s, when the federal government supported national projects on "agricultural modernisation" that were derived from agreements with the Ford and Rockefeller foundations, the main campaigners of the Green Revolution. A strategy proposed by the workshop participants is associated with community governance:

> Here we realise the problems, we are talking about it, and they are problems that we have in common, even if we come from different places. We have to go now to talk to the people in our communities, we must understand that transgenic maize and pesticides are not the best options. We need a lot of organisation and communication to change those beliefs.

Regarding soil and water contamination, it was mentioned that the problem was not only the application of chemicals in *milpas*, but also the mismanagement of pesticide containers. One suggested strategy to address this problem is again related to community governance:

> In my community, we are making citizen complaints at all three levels of government (municipal, state and federal). We also have been able to make alliances with other organisations and universities ... we cannot continue to think that this is God's work— this is because there is bad ecosystem management.

Another proposal from participants involved innovation. They noted that it is essential for each community to find a way to get water filters, and also to get technologies for rain capture.

A proposal on addressing the shortcomings of health services in communities involved the creation of medicine hospitals:

> Young people could learn traditional medicine and train others so that we will no longer have to go looking for hospitals that are far away. We have to look for alternatives together.

Another of the strategies mentioned was to seek advice on how to know their rights with regard to human health in order to be able to demand them.

4 Discussion

The livelihoods and health of FML inhabitants are closely related to the proper functioning of ecosystems. The rainforest is essential to maintain the hydrological cycle in the FML, which means that rains preserve *milpas* and provide water for human consumption. Water is also used for irrigation of home gardens, to feed animals, and to maintain apiaries. In turn, all production activities conducted in the soil of the FML have a substantial impact on water quality due to the soil's high permeability. According to Batllori (2017), the main sources of water pollution include all those mentioned by the participants, including pig farms, agricultural activities, and garbage dumps, but also wastewater from tourist areas and urban areas that, although seem to be non-rural issues, have a significant impact on FML water quality.

The recognition of the close relationship between human health and ecosystem health has been lost over time. Older people continue to perform rituals of respect and appreciation for what nature provides to them (e.g. food, medicine, beauty), but younger generations have left these customs behind. The devalorisation of cultural ecosystem services in SEPLS has been identified as one of the main fronts of scholarly enquiry in the new geographies of conservation (Sarmiento & Cotacachi, 2019). Duarte (2017) reflected on the change in perceptions of resources, such as water or maize, noting that perceptions define how resources are valued and handled. She also recounts how on a visit to a *J'men* (a man specialising in the rituals of the Mayan *milpa*), she heard him say, "the holy water is a blessing to all human beings living on earth" (Duarte, 2017, p. 137). In this sense, the proposals made during the workshops on recovering and revaluing local knowledge by creating traditional medicine hospitals are very relevant, not only to redevelop and strengthen the meaning of the "human health-nature" relationship, but also to address a violation of a fundamental human right—access to public health systems.

Food production is another relevant issue for FML health. Some studies have analysed the relationship of pesticides with diseases, such as cancer, showing that they inhibit humans' hormonal functions (Polanco, López, et al., 2018b). Pesticides enter the FML's food chains from various sources, including honey, *milpa* products, bush meat, and water. Participants' proposals to develop sustainable innovations for production practices and access to less contaminated water may fail to eliminate many pollution sources; however, these proposals are essential to ensure that production activities for self-consumption, like beekeeping and the *milpa* system, remain healthy systems for food security, and no longer function as another vehicle for pollutants to enter the human body.

Although our study is based on an analysis of the perceptions of FML inhabitants, it is interesting that their perceptions are closely associated with information from the scientific literature. For example, the concern they expressed in the workshops about water pollution associated with pig farms and pesticides has been documented in various studies previously mentioned in this chapter (Batllori, 2017; Hernández & Ortega, 2017; Metcalfe et al., 2011; Polanco et al., 2017; Polanco, Araujo, et al., 2018a). The problem of diabetes being associated with changes in diet has also been widely documented, and is relevant for these communities whose diet was previously based on the consumption of *milpa* products. For example, Narvaez and Segura (2020) analyse the recent increase of diabetes in indigenous Mayan regions; they also explain the antidiabetic potential of foodstuff from the *milpa*. Some studies have documented evidence showing that the introduction of industrialised foods into the diet of Mayan communities a decade ago, which meant a less nutritious diet, rich in sugars, saturated fats, salt, and colour and flavour additives, implied an increased risk of obesity and diabetes (Pérez Izquierdo et al., 2012; Leatherman & Goodman, 2005). In recent studies this problem prevails (Frank & Durden, 2017; Leatherman et al., 2020; Otero Prevost et al., 2017). Although the subject of dietary change related to gender did not arise in the workshops, it is relevant. For instance, obesity is a problem that affects women more in Mayan communities than elsewhere (Marín Cárdenas et al., 2014).

The generation of resilience indicators revealed the many vulnerabilities of the FML. However, it also allowed reflection on potential options to seek to enhance the landscape's resilience from the local point of view, accepting that the cause of vulnerabilities often comes from decisions made outside the FML communities. One of the most important lessons of this exercise is that assessing trends, reflecting collectively, and designing strategies from a bottom-up perspective have the potential to empower local people to begin to recognise themselves as managers of their territories.

5 Conclusion

One main lesson learned is that the landscape approach methodology is much more aligned with the peasants' view of their ecosystem than a sectoral approach. During the feedback sessions with participants, they mentioned that the integrated approach, for example including medicine produced within the ecosystem, allowed them to relate their experiences to the future SGP strategy. In rich biocultural diverse contexts, this approach is even more relevant.

The resilience indicators, co-created by the participants, focused on the perception of trends. In the development context, data is the main source of information to define a baseline. By asking participants how they visualise those trends, we recognise that they own the key knowledge. The methodology also allows for the detection of threats that are not monitored by governmental bodies, such as the use of pesticide in the rural sector in Mexico.

In 2019, during the workshops, the human health indicator was chosen as a priority by the participants. Because the strategies proposed are not GEF-eligible activities, adoption by the SGP of a specific strategy was considered as a challenge by the country's programme team. However, one year later, and after 14 months of the SARS-CoV-2 pandemic, the GEF is now holding discussions within its Scientific and Technical Advisory Panel on the link between ecosystem and human health (Scientific and Technical Advisory Panel (STAP), 2020). To add to these discussions, new theoretical proposals on the emergence of SARS-CoV-2 on the planet should be a priority. Alcántara-Ayala (2021) proposes the typification of this type of phenomenon as "syndemic pan-disaster". This approach recognises that society's vulnerability and exposure to COVID-19 have presented the great challenge of solving countless pre-existing problems, such as those detected in the exercise documented in this chapter. Thus, the communities detected local threats that are now fully part of the global environmental agenda, and action may be taken during the next 10 years in SGP Mexico to support those initiatives in the framework of the United Nations Decade on Ecosystem Restoration.

Finally, it is essential to mention that almost half of the participants in the workshops were women, highlighting the importance of the gender perspective during these workshops. This topic becomes highly relevant when scientific data show that environmental health affects men and women differently, for example, the relationship between uterine and breast cancer, and contaminated breast milk and pesticide contamination in water (Polanco et al., 2017; Polanco, López, et al., 2018b). The gender perspective makes it possible to generate strategies that better address the problems faced by SEPLS.

Acknowledgements This project was funded by SGP-GEF. We want to thank the Mayan men and women who shared with us their reflections and their knowledge.

References

Aide, T. M., Clark, M. L., Grau, H. R., López-Carr, D., Levy, M. A., Redo, D., Bonilla-Moheno, M., Riner, G., Andrade-Núñez, M. J., & Muñiz, M. (2013). Deforestation and reforestation of Latin America and the Caribbean (2001–2010). *Biotropica, 45*(2), 262–271.

Alcántara-Ayala, I. (2021). COVID-19, más allá del virus: una aproximación a la anatomía de un pandesastre sindémico. Investigaciones Geográficas, 104. https://doi.org/10.14350/rig.60218

Barrera-Bassols, N., & Toledo, V. M. (2005). Ethnoecology of the Yucatec Maya: Symbolism, knowledge and management of natural resources. *Journal of Latin American Geography, 4*, 9–40.

Batllori, E. (2017). Condiciones actuales del agua en la Península de Yucatán. In M. Chávez-Guzmán (Ed.), *El manejo del agua a través del tiempo en la Península de Yucatán* (pp. 201–225). UADY. isbn:978-607-8527-12-0.

Bergamini, N., Dunbar, W., Eyzaguirre, P., Ichikawa, K., Matsumoto, I., Mijatovic, D., Morimoto, Y., Remple, N., Salvemini, D., Suzuki, W., & Vernooy, R. (2014). *Toolkit for the indicators of resilience in socio-ecological production landscapes and seascapes*. UNU-IAS, Biodiversity International, IGES & UNDP.

Comisión Nacional para el Conocimiento y Uso de la Biodiversidad (CONABIO). (2015). Límites y regionalización de los Corredores Biológicos del sureste de México, 2015, escala: 1:250000, México, D. F.

Comisión Nacional de Áreas Naturales Protegidas. (2020). viewed 14 December 2020. Retrieved from https://www.gob.mx/conanp/documentos/region-peninsula-de-yucatan-y-caribe-mexicano?state=published.

Duarte, A. (2017). La "santa agua" en la cutura maya en Yucatán. In M. Chávez-Guzmán (Ed.), *El manejo del agua a través del tiempo en la Península de Yucatán* (pp. 137–151). UADY. isbn:978-607-8527-12-0.

Frank, S. M., & Durden, T. E. (2017). Two approaches, one problem: Cultural constructions of type II diabetes in an indigenous community in Yucatán, Mexico. *Social Science & Medicine, 172*, 64–71.

García-Frapolli, E., Toledo, V. M., & Martínez-Alier, J. (2008). Apropiación de la naturaleza por una comunidad maya yucateca: un análisis económico-ecológico. *REVIBEC-Revista Iberoamericana de Economia Ecologica, 7*, 27–42.

Gutiérrez Núñez, N. L. (2020). Revolución verde en los suelos agrícolas de México. Ciencia, políticas públicas y agricultura del maíz, 1943-1961. *Mundo Agrario, 21*(47), e142. https://doi.org/10.24215/15155994e142

Hernández, L., & Ortega, D. (2017). El agua en la Península de Yucatán. In M. Chávez-Guzmán (Ed.), *El manejo del agua a través del tiempo en la Península de Yucatán* (pp. 37–45). UADY. isbn:978-607-8527-12-0.

Instituto Nacional de Pueblos Indígenas. (2015). *Atlas de los pueblos indígenas de México*, viewed 5 December 2021. Retrieved from http://atlas.inpi.gob.mx/?page_id=7256.

Instituto Nacional de Estadística y Geografía (INEGI). (2015a). *Población Económicamente Activa de México*.

Instituto Nacional de Estadística y Geografía (INEGI). (2015b). *Encuesta intercensal: Lengua Indígena*.

Instituto Nacional de Estadística y Geografía (INEGI). (2016). Conjunto de datos vectoriales de uso de suelo y vegetación, Escala 1:250,000, Serie VI.

Leatherman, T. L., & Goodman, A. (2005). Coca-colonization of diets in the Yucatan. *Social Science & Medicine, 61*(4), 833–846.

Leatherman, T., Goodman, A. H., & Stillman, J. T. (2020). A critical biocultural perspective on tourism and the nutrition transition in the Yucatan. In H. Azcorra & F. Dickinson (Eds.), *Culture, environment and health in the Yucatan Peninsula* (pp. 97–120). Springer.

Mariaca, R. (2015). La milpa maya yucateca en el siglo XVI: evidencias etnohistóricas y conjeturas. *Etnobiología, 13*(1), 1–25.

Marín Cárdenas, A. D., Sánchez Ramírez, G., & Maza Rodríguez, L. L. (2014). Prevalencia de obesidad y hábitos alimentarios desde el enfoque de género: el caso de Dzutóh, Yucatán, México. *Estudios Sociales, 22*(44), 64–90.

Martínez, F., Benítez, M., Ramos, X., García, G., Bracamontes, L., & Vázquez, B. (2017). *Derechos humanos y patrimonio biocultural. El sistema milpa como cimiento de una política de estado cultural y ambientalmente sustentable*. Centro Mexicano de Derecho Ambiental (CEMDA).

Metcalfe, C. D., Beddows, P. A., Bouchot, G. G., Metcalfe, T. L., Li, H., & Van Lavieren, H. (2011). Contaminants in the coastal karst aquifer system along the Caribbean coast of the Yucatan Peninsula, Mexico. *Environmental Pollution, 159*(4), 991–997.

Narvaez, J. J., & Segura, M. R. (2020). Foods from Mayan communities of Yucatan as nutritional alternative for diabetes prevention. *Journal of Medicinal Food, 23*(4), 349–357.

Odum, E. P., & Sarmiento, F. O. 1998, *Ecología: El Puente Entre Ciencia y Sociedad*, Editorial MacGraw Hill-Interamericana de Mexico.

Otero Prevost, D. E., Delfín Gurri, F., Mariaca Méndez, R., & Guízar Vázquez, F. (2017). La incorporación y el aumento de oferta de alimentos industrializados en las dietas de las unidades domésticas y su relación con el abandono del sistema de subsistencia propio en las comunidades rurales mayas de Yucatán, México. *Cuadernos de Desarrollo Rural, 14*(80), 1–16.

Pérez Izquierdo, O., Nazar Beutelspacher, A., Salvatierra Izaba, B., Pérez-Gil Romo, S. E., Rodríguez, L., Castillo Burguete, M. T., & Mariaca Méndez, R. (2012). Frecuencia del consumo de alimentos industrializados modernos en la dieta habitual de comunidades mayas de Yucatán, México. *Estudios Sociales (Hermosillo Son), 20*(39), 155–184.

Pinto, L. S., & Barrios, L. G. (2015). Conservación en espacios rurales humanizados, Ecofronteras.

Polanco, A. G., López, M. I. R., Casillas, T. A. D., León, J. A. A., Prusty, B. A. K., & Cervera, F. J. Á. (2017). Levels of persistent organic pollutants in breast milk of Maya women in Yucatan, Mexico. *Environmental Monitoring and Assessment, 189*(2), 59.

Polanco, A. G., Araujo, A., Tamayo, J. M., & Munguía, A. (2018a). The glyphosate herbicide in Yucatan, Mexico. *MOJ Bioequivalence & Bioavailability, 5*(6), 284–286. https://doi.org/10.15406/mojbb.2018.05.00115

Polanco, A. G., López, M. I. R., Casillas, Á. D., León, J. A. A., & Banik, S. D. (2018b). Impact of pesticides in karst groundwater. Review of recent trends in Yucatan, Mexico. *Groundwater for Sustainable Development, 7*, 20–29.

Rodríguez-Robayo, K. J., Méndez-López, M. E., Molina-Villegas, A., & Juárez, L. (2020). What do we talk about when we talk about milpa? A conceptual approach to the significance, topics of research and impact of the Mayan milpa system. *Journal of Rural Studies, 77*, 47–54.

Salazar Barrientos, L. L., & Magaña Magaña, M. Á. (2016). Aportación de la milpa y traspatio a la autosuficiencia alimentaria en comunidades mayas de Yucatán. *Estudios Sociales (Hermosillo Son.), 24–25*(47), 182–203.

Salazar Barrientos, L. L., Magaña Magaña, M. Á., Aguilar-Jiménez, A. N., & Ricalde-Pérez, M. F. (2016). Factores socioeconómicos asociados al aprovechamiento de la agrobiodiversidad de la milpa en Yucatán. *Ecosistemas y recursos agropecuarios, 3*(9), 391–400.

Sarmiento, F. O., & Cotacachi, C. (2019). Framing cultural ecosystem services in the Andes: Utawallu Runakuna as sentinels of values for biocultural heritage conservation. In UNU-IAS & IGES (Ed.), *Understanding the multiple values associated with sustainable use in socio-ecological production landscapes and seascapes (SEPLS)* (Satoyama initiative thematic review) (Vol. 5, pp. 31–46). United Nations University Institute for the Advanced Study of Sustainability.

Scientific and Technical Advisory Panel (STAP). (2020). STAP Chair's Report to the GEF Council (pp. 1–16).

Terán, S. (2010). Milpa, biodiversidad y diversidad cultural. Contexto social y económico. In R. Durán & M. Méndez (Eds.), *Biodiversidad y Desarrollo Humano en Yucatán*. CICY, PPD-FMAM, CONABIO, SEDUMA.

Toledo, V., Ortíz-Espejel, B. F., Cortés, L., Moguel, P., & Ordonez, M. D. J. (2003). The multiple use of tropical forest by indigenous people in México: A case of adaptive management. *Conservation Ecology, 7*(3), 9.

Toledo, V. M., Barrera-Bassols, N., García-Frapolli, E., & Alarcón-Chaires, P. (2008). Multiple use and biodiversity within the Mayan communities of Yucatan, Mexico. *Interciencia, 33*(5), 345–352.

Toledo, V. M., & Barrera-Bassols, N. (2011). Saberes tradicionales y adaptaciones ecológicas en siete regiones indígenas de México. In F. R. Escutia & S. B. Garcia (Eds.), *Saberes ambientales campesinos: Cultura y naturaleza en comunidades indígenas y mestizas de México* (pp. 15–60). Universidad de Ciencias y Artes Chiapas.

Zizumbo-Villarreal, D., & Colunga-García, P. (2017). La milpa del occidente de Mesoamérica: profundidad histórica, dinámica evolutiva y rutas de dispersión a Suramérica. *Revista de Geografía Agrícola, 58*, 33–46.

The opinions expressed in this chapter are those of the author(s) and do not necessarily reflect the views of UNU-IAS, its Board of Directors, or the countries they represent.

Open Access This chapter is licenced under the terms of the Creative Commons Attribution 3.0 IGO Licence (http://creativecommons.org/licenses/by/3.0/igo/), which permits use, sharing, adaptation, distribution and reproduction in any medium or format, as long as you give appropriate credit to UNU-IAS, provide a link to the Creative Commons licence and indicate if changes were made.

The use of the UNU-IAS name and logo, shall be subject to a separate written licence agreement between UNU-IAS and the user and is not authorised as part of this CC BY 3.0 IGO licence. Note that the link provided above includes additional terms and conditions of the licence.

The images or other third party material in this chapter are included in the chapter's Creative Commons licence, unless indicated otherwise in a credit line to the material. If material is not included in the chapter's Creative Commons licence and your intended use is not permitted by statutory regulation or exceeds the permitted use, you will need to obtain permission directly from the copyright holder.

Chapter 6
Community "Bio-Rights" in Augmenting Health and Climate Resilience of a Socio-Ecological Production Landscape in Peri-urban Ramsar Wetlands

Dipayan Dey and Priyani H. Amerasinghe

Abstract In the climate milieu, peri-urban wetlands are facing the serious threats of habitat destruction, biodiversity loss, and deteriorating ecosystem services owing to anthropogenic pressure and rapidly changing microclimates. Although some of these wetlands are unique socio-ecological production landscapes and seascapes (SEPLS) that ensure the food, water, and livelihood security of urban poor, they remain excluded from mainstream conservation. Ecosystem-based adaptive conservation and wise use by communities are sustainable solutions to protect these SEPLS, wherein the opportunity costs of wetland conservation to the ultra-poor are compensated with payments for ecosystem services. This chapter documents the success of a rights-based, neo-economic conservation model, entitled "*Bio-rights of commons*", in two such peri-urban Ramsar wetlands, the East Kolkata Wetlands (EKW) and the Deepor Beel Wetland (DBW), both on the brink of extinction. The bio-rights model was developed by the South Asian Forum for Environment (SAFE) under the aegis of the Ramsar Secretariat in 2010 and implemented in the East Kolkata Ramsar wetlands. Perusal of results revealed that in both SEPLS, a rights-based conservation approach could ensure livelihood security as well as health and well-being during post-pandemic stress. A circular economic intervention was enabled at the community-ecosystem interface, through capacity-building in wastewater-fed captive fisheries, ecotourism in wetlands, and organic waste recycling as alternative livelihood opportunities. This compensated for the opportunity costs incurred by the wetland communities in conserving the SEPLS and also ensured community "bio-rights" to the wetlands' ecosystem services. While these efforts steadied biodiversity indices and waterbody permanence of these Ramsar wetlands, they also provided fresh air for the pollution-wracked cities of Kolkata and Guwahati during the

D. Dey (✉)
South Asian Forum for Environment, Kolkata, West Bengal, India
e-mail: chair@safeinch.org

P. H. Amerasinghe
International Water Management Institute, Colombo, Sri Lanka

© The Author(s) 2022
M. Nishi et al. (eds.), *Biodiversity-Health-Sustainability Nexus in Socio-Ecological Production Landscapes and Seascapes (SEPLS)*, Satoyama Initiative Thematic Review, https://doi.org/10.1007/978-981-16-9893-4_6

COVID-19 pandemic, and augmented economic opportunities in fisheries for land-less casual labourers migrating back home during the countrywide lockdown. The intervention recommended an operational guideline for policy frameworks in sustainably conserving these wetland SEPLS for enriching biodiversity, human health, and well-being.

Keywords Bio-rights · Peri-urban wetlands · Ecosystem-based conservation · Ecosystem services

1 Introduction

Wetlands provide a plethora of nature's goods and services, constituting unique socio-ecological production landscapes and seascapes (SEPLS) that ensure food, water, and livelihood security, especially for the poor. However, Osinuga and Oyegoke (2019) observed that these SEPLS are still excluded from mainstream conservation and are facing serious threats owing to anthropogenic pressure in the milieu of climate change. This has led to habitat destruction, biodiversity loss, and deterioration of ecosystem services, widening the poverty trap. Bassi et al. (2014) acknowledge that Indian peri-urban wetlands are degrading faster due to both biotic and abiotic threats like drainage and landfill, over-exploitation of fish resources, discharge of industrial effluents, uncontrolled siltation, and weed infestation, and the ill-effects of fertilisers, pesticides, and detergents. Reports in the India Water Portal reveal that the South Asian ecoregion is losing 25 km^2 wetlands per square kilometre of urban encroachment (Bansal, 2020), a rate even faster than the loss of forest cover (Reddy et al., 2018). One-third of wetlands in the Indian subcontinent have already been severely degraded, warranting policies for the management of this critical ecosystem. In the context of climate change, the mitigation potentials of wetlands are very high, though urban planning for the conservation of wetlands in smart cities is meagrely addressed (Mcinnes, 2014).

This chapter documents the success of a rights-based, neo-economic conservation model, entitled "*Bio-rights of Commons*" in two such peri-urban Ramsar wetlands of eastern India, the East Kolkata Wetlands (EKW) and the Deepor Beel Wetland (DBW). These wetlands were on the brink of being removed from the Ramsar list owing to deteriorating habitat health and biodiversity loss (Ghosh & Das, 2020; Saikia, 2019). After successful implementation of the bio-rights project, both wetlands are able to better support the livelihoods of the communities dependent on them for ecosystem services (ESs). The project has contributed to wetland habitat restoration and prevention of biodiversity loss, and simultaneously, sustainable intensification of ESs has substantially contributed to health benefits for the communities during the present pandemic.

EKW is a multifunctional sewage-fed wetland complex, spanning a 12,741-hectare area in the Bhagirathi-Ganga riverine floodplain that naturally recycles nearly 250 million gallons of sewer water and removes around 237 kg of biological oxygen demand (BOD) per day (Bhattacharyya et al., 2012). It was recognised as the world's largest natural resource recovery ecosystem in the fringe of Kolkata Metropolis by the Ramsar Secretariat in 2002. Until 1830, tidal flow from the Bay of Bengal through rivers and rivulets drenched this area with saline water; but in the latter half of the nineteenth century, as the rivers dried out due to siltation, the area started receiving urban wastewater and local communities began wastewater fish farming (Ghosh & Sen, 1987). Since then, local indigenous communities have been traditionally managing the sewer flow to regulate the nutrient load, thereby leveraging the natural resource recovery system of EKW for intensive fisheries and organic farming (Ghosh, 2005). EKW provides an array of ecosystem services to support livelihoods and the wetland environment relating to provisioning of wastewater recycling, food and fish production, flood protection, and more. It also harbours a rich biodiversity of fishes, birds, reptiles, amphibians, and plants (Ghosh, 1990).

DBW is an oxbow lake of the Brahmaputra River, situated in close proximity to the city of Guwahati. It is a typical wetland ecosystem within the Burma Monsoon Forest Biogeography region that provides various goods and services to the local community and animal population of the Rani-Garbhanga reserved forests, which are adjacent to the wetlands and house nearly 120 Asiatic elephants (Mitra & Bezbaruah, 2014). DBW is traditionally managed as a major fish breeding and nursery ground by three indigenous fishing communities for supplying fish stocks to other nearby waterbodies while conserving the local fish biodiversity. Almost 750 households in 14 fringe villages are directly dependent on the wetland for fishing and collection of herbaceous plants (Mahadevia et al., 2017).

Both of these wetlands are declared Ramsar sites. Biophysical and socio-economic studies have been carried out, and management plans for the wetlands have been proposed. Yet, the current approach to management appears fragmented, and innovative ways are needed to conserve and protect the ecosystem services of the wetlands (Mukherjee, 2011). A thorough institutional analysis and responses to demands arising from impacts on wetland ecosystem services due to urban expansion are needed. Moreover, a large number of poor who depend on the ecosystem services of the wetlands are still excluded from the decision-making processes of these SEPLS (Dey, 2008).

This chapter explores a rights-based approach to poverty alleviation and SEPLS conservation. It reviews bio-rights as a community-based neo-fiscal conservation paradigm for protecting ecosystems of global ecological importance by compensating the opportunity cost incurred by the communities living in the vicinity and dependent on the environmental goods and services of the ecosystem. The global average compensation cost would not be difficult to accomplish if technological cooperation and developmental resources were equitably extended to such communities. Bio-rights is therefore a sustainable neo-financing mechanism where the poor can be empowered to protect the ecosystem services that they depend upon. It is envisaged that a better institutional management framework that includes bio-rights can be developed to safeguard the ecosystem services of the wetlands, facing the menace of urbanisation.

2 Methodology

2.1 Intervention Area

The major study was conducted in 2019 in two wetland areas, namely the East
Kolkata Wetlands (22.55°N 88.45°E) and the Deepor Beel Wetland (26.13°N
91.66°E) (Figs. 6.1 and 6.2 and Table 6.1). The EKW is a complex of natural and
man-made wetlands lying east of the city of Kolkata, spreading over the districts of
North and South 24 Parganas in the state of West Bengal, India. The wetlands cover
12,741 hectares at an elevation of 12 m above sea level and constitute one of the
largest assemblages of sewage-fed fish ponds including salt marshes, agricultural
fields, and settling ponds. The wetlands naturally treat the city's sewage, with a
detention period of 35 days, and the nutrients contained in the wastewater sustain
fish farms and agriculture in 264 fishery cooperatives.

Deepor Beel is located at an elevation of 53 m above sea level to the southwest of
Guwahati city, in the Kamrup district in the state of Assam, India. It is a permanent
freshwater oxbow lake, covering 4014 hectares with an average depth of 1.5 ft., that
originated from a former channel to the south of the Brahmaputra River. The wetland
complex is within close vicinity of Guwahati and fragmented by highways and
railroads. It was recognised as a Ramsar site owing to its vast biodiversity and

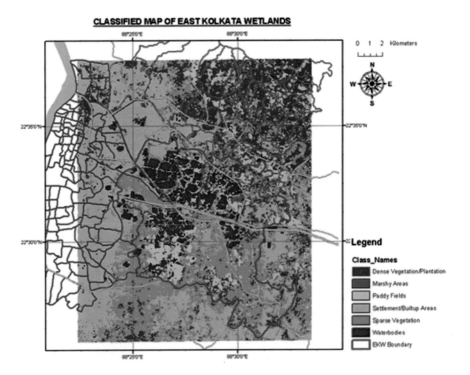

Fig. 6.1 Classified map of East Kolkata Wetlands (source: prepared by author)

Fig. 6.2 Location map of Deepor Beel Wetland (source: prepared by author)

Table 6.1 Basic information of the study area

Country	India	India
Province	West Bengal	Assam
District	Kolkata	Kamrup Urban
Size of geographical area (hectare)	12,500	4014
Number of direct beneficiaries (persons)	1600	1400
Number of indirect beneficiaries (persons)	145,000	45,000
Dominant ethnicity(ies), if appropriate	Indigenous fishers (*Sardar, Koiri*)	Indigenous fishers (*Karbi, Koiborto*)
Size of the case study/project area (hectare)	125	500
Geographic coordinates (latitude, longitude)	22.5528°N, 88.4501°E	26.13°N 91.66°E

ecosystem services that sustain the livelihoods of nearly 1200 households of indigenous communities.

2.2   Geospatial Mapping

Three land-use pattern and land cover (LUP-LC) maps of both peri-urban wetlands were prepared using Q-GIS open-access mapping software and based on satellite images of the base year 2000 and years 2009 and 2019. Maps were analysed and compared to identify both major and minor land-use areas for change detection studies. Following this, random physical surveys and GPS ground truthing were conducted in these areas to validate the GIS maps as well as the ecosystem services of the wetlands. Areas having wetlands that serve agriculture were identified and overlayed on land-use change detection maps. Inferences were drawn on changes in LUP-LC and subsequent changes in the ESs based on the study results.

2.3   Biodiversity and Ecosystem Service Assessments

The Trophic State Index (TSI), a frequently used biomass-related index (Carlson, 1977) that ranges between 60 and 100 on average, was used to estimate the water condition and biodiversity of the waterbodies. The TSI of phosphorus, TSI of Secchi disc depth, and TSI of chlorophyll-a were averaged to arrive at a single index ranging between 60 and 100. Water samples for the analysis of phosphorus and chlorophyll-a were collected from the waterbodies using subsampling methods to form one composite sample from each site (Davies & Tsomides, 2014). Four subsamples of 500 mL each were transferred to a churn spitter and the composite sample was transferred to a 2 L sampling bottle. The bottles were previously cleaned with a detergent devoid of phosphorus, washed with acid, and rinsed vigorously thrice with distilled water (American Public Health Association (APHA), 1995; U.S. Geological Survey (USGS), 2006). For phosphorus sampling the water sample was subjected to perforated digestion followed by ascorbic acid method, 4500E (American Public Health Association (APHA), 1995). Chlorophyll-a was estimated by acetone method. A spectrophotometer was used for analysis. Absorbance was marked at 630 nm (APHA 1995). The units for both chlorophyll-a and phosphorus are microgram/litre. Equations for TSI are as follows:

TSI(SD) $= 60 - 14.41 \ln (SD)$.

TSI(CHL) $= 9.81 \ln (CHL) + 30.6$.

TSI(TP) $= 14.42 \ln (TP) + 4.15$.

Note: TSI is Carlson Trophic State Index, SD is Secchi disc, CHL is chlorophyll, and TP is total phosphorus

Ecosystem services (ESs) were classified into four major categories, namely, (1) provisioning services, (2) regulating services, (3) habitat and supporting services, and (4) cultural services, and were assessed following the TEEB list (de Groot et al., 2010). One of the major objectives was to identify the existing ESs of the peri-urban wetlands, as well as the degree of livelihood dependency on these wetland ESs, with special weightage given to the use of wetlands for agriculture. Surveys of the ESs of the peri-urban wetlands were conducted on 104 waterbodies throughout EKW and DBW, of which 43 support agricultural practices like aquafarming and horticulture. The survey was done by stratified sampling of users of ESs, based on gender, age groups, and livelihood vulnerability index (LVI) scores with a preset questionnaire. Focus group discussions and rapid rural appraisal (RRA) for valuation of ESs were also conducted following standard protocols. Likewise, to identify the livelihood dependency of the peri-urban communities, gender-balanced focus group discussions with 12–15 members were conducted for both the EKW and DBW wetland areas.

2.4 Production Analysis

To estimate and analyse the commercial productivity of the wetlands, spatio-temporal measurements of fish yield were taken in metric tons per hectare every month for both intensive and non-intensive farming practices. Yield data on fish production were collected pre-monsoon, during monsoon, and post-monsoon from six different locations in both the EKW and DBW, and used to determine the mean value and provide trend analysis. Differential data on self-consumption and sale was also collected.

2.5 Sociometric Indexing

The sociometric study is based on primary questionnaire surveys in the peri-urban agricultural areas of Kolkata for EKW and Guwahati for DBW. The target group was families who depend on agriculture for their livelihood. The LVI scores were based on eight major components: socio-demographic profile, livelihood strategy, health, natural capital, social network, food, water, and natural hazards and climate vulnerability (Hahn et al., 2009). Each is comprised of various subcomponents. The LVI uses a balanced weighted average approach where each subcomponent contributes equally to the overall index, even though each major component is comprised of a different number of subcomponents. Because each of the subcomponents is measured on a different scale, each component is indexed first. The steps for calculation included the following:

(a) Balanced weighted average approach where each subcomponent contributes equally to the overall index. As each of the subcomponents is measured on a different scale, it is necessary to standardise each as an index. For this, the following equation is used:

$$Index_{sd} = Sd - Smin/Smax - Smin,$$ where Sd is the original value of the subcomponent and $Smin$ and $Smax$ are the minimum and maximum values in the study area.

(b) After standardising each subcomponent, the subcomponents are averaged using the following equation:

$$Md = \sum Index_{sd/n},$$ where Md = one of the eight major components and $Index_{sd/n}$ represents the subcomponents that make up each major component.

(c) Once the values of each of the eight major components are calculated, they are averaged to obtain LVI using the following equation: $LVI = \sum W_M.Md/\sum W_M$, where W_M is the number of subcomponents in each major component, and LVI equals the weighted average of the eight major components. The range of LVI data is always between 0 and 1.

2.6 Attitude Scaling

In order to quantify perceptions and levels of awareness on nature's goods and services, as well as their significance for sustainable development and livelihood security among the beneficiaries, changes in attitude towards conservation priorities, wise use of wetlands, and overall vulnerability were measured using a five-point Likert scale.

3 Results and Outputs

3.1 Urban Encroachments and LUP-LC Changes

The land-use maps show that there has been significant loss of peri-urban wetland area in both sites impacting the livelihood patterns of the local communities. The results in EKW show an alarming loss amounting to 53.28% of the total wetland area over two decades of time, mainly due to shrinking of wetland boundaries as well as conversion of agricultural land (paddy fields) to forest-like cover (orchards and trees) by 27.44%. However, there has been a nominal increase of 2.7% in built-up areas and fallows compared to the base year 2000. In the DBW complex, wetland loss was 47%, with a 32% increase in built-up areas and waste dumping areas. Field observations showed that owing to loss of wetlands, there has been a shortage in water supply, as ponds and drainage were clogged leading to conversion of agricultural lands to small fragments of horticultural plots. Intensive horticulture was promoted in the fertile land residues as fisheries and paddy production declined. This was furthered by intensive chemical farming for yield enhancement, and later these plots were ultimately converted into orchards and plantations. This land-use change has been more drastic on the urban fringes in both EKW (Fig. 6.3) and DBW.

3.2 Biodiversity and Ecosystem Services

The difference between monsoon and post-monsoon TSI (chlorophyll) was remarkable in EKW, with the eutrophication level during the post-monsoon period surpassing that of monsoon period. In DBW, this phenomenon is reversed, and the monsoon TSI is higher than the post-monsoon TSI. Thus, even during the monsoon when the volume of water flow is higher, the waterbodies stay eutrophicated. This feature in DBW can be explained by the clogged canal system, which serves as both outlet and inlet for this particular SEPLS. The inflow and outflow of water from the waterbody are staggered due to habitat fragmentations. This in turn increases algal mass and phytoplankton growth at a rate not discerned in EKW, wherein sewer water drainage (both inlet and outlet) is traditionally controlled and regulated by the

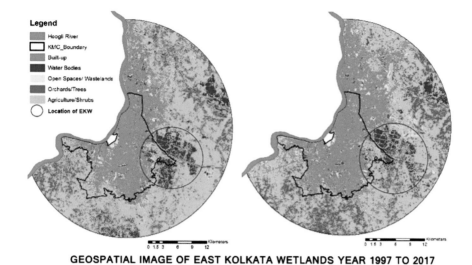

Legend
- Hoogli River
- KMC_Boundary
- Built-up
- Water Bodies
- Open Spaces/ Wastelands
- Orchards/Trees
- Agriculture/Shrubs
- ○ Location of EKW

GEOSPATIAL IMAGE OF EAST KOLKATA WETLANDS YEAR 1997 TO 2017

Fig. 6.3 LUP-LC changes in the peri-urban East Kolkata Wetlands (source: Amerasinghe & Dey, 2018)

communities. The post-monsoon TSI levels in EKW are higher than the monsoon TSI levels, showing betterment in the health of the trophic status of the waterbodies in the monsoon period. This is attributed to the diluted nutrients as well as the free flow of excess water during the monsoon period reducing biomass productivity and leading to lower levels of algal bloom and phytoplankton growth.

The average TSI levels of all the waterbodies, in general, are quite high (>80) in both the monsoon and post-monsoon period. This leads to the inference that all the waterbodies are hypereutrophic, which suggests a high production capacity as a result of excessive nutrients with visibility less than 15 cm on a Secchi disc. This leads to excessive growth of algal scum and dominance of bottom-dwelling preda-tory fish species, and prevents aquatic life from functioning at lower depths, creating dead zones in the subsurfaces, resulting in loss of biodiversity and shortening of trophic levels. Waterbodies in the area in EKW, managed by communities under the bio-rights conservation programme, show borderline eutrophication and a relatively low post-monsoon TSI and a lower nutrient level compared to the other waterbodies, wherein there is a low chance of fish death and hyper-eutrophication as well. Table 6.2 shows some TSI data for substantiating the findings.

The study showed that both wetland SEPLS provide regulating services such as maintenance of local climate and air quality, carbon sequestration, moderation of extreme events such as floods and drought, erosion prevention, maintenance of soil quality, and pollination. Home to a huge number of species, these wetlands help to maintain genetic diversity, and thus also provide habitat and supporting services. Of the four categories of ESs, provisioning services are the most prominent as they directly serve the basic needs of the local community, as well as provide livelihood support to a large population. Major provisioning services include providing space

Table 6.2 Seasonal TSI (chlorophyll and Secchi disc) in EKW and DBW

Location	Site no.	Monsoon			Post-monsoon		
		TSI (Chl)	TSI (SD)	TSI (Av)	TSI (Chl)	TSI (SD)	TSI (Av)
EKW	EK1	71.351882	81.552998	80.3618978	88.5646825	93.18025	82.5570143
	EK2	74.30124	84.552998	81.887169	85.9789184	89.39958	83.083291
DBW	DB1	81.033981	91.806832	85.9066857	80.1892666	89.39958	84.4527717
	DB2	83.274913	88.331686	86.6699869	82.3475607	87.33752	87.6235635

Major Ecosystem Services in Peri-urban Wetlands

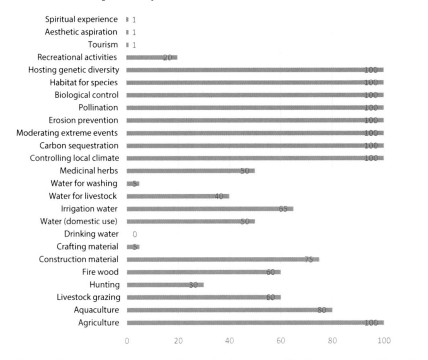

Fig. 6.4 Major ecosystem services of the peri-urban waterbodies (source: prepared by authors)

for agriculture, aquaculture, livestock grazing, and hunting; providing raw materials for firewood, construction, and crafts; providing fresh water for domestic use, irrigation, washing livestock, drinking water for livestock, and commercial washing; and providing medicinal herbs. Land for agriculture and aquaculture are the two most important ecosystem services, followed by water for irrigation, land for livestock grazing, and water for domestic usage. Although cultural services very much exist in the peri-urban wetlands, they are not clearly conceived by the local community (Fig. 6.4).

The results from sociometric studies and FGDs showed that the ESs of the peri-urban wetlands are of huge importance to the peri-urban community. They fulfil the basic needs of the local community as well as serve as a major source of income generation and livelihood. Services such as land for agriculture, aquaculture, and livestock grazing; collection of materials for crafts; and extraction of water for irrigation, livestock washing, and commercial washing provide support for income generation. On the other hand, collection of firewood, construction material, extraction of fresh water for domestic use, collection of medicinal herbs, and collection of food through hunting and gathering fulfil the basic needs of the local people. Aquaculture is the most important livelihood supporting ecosystem service of the peri-urban wetlands in Kolkata, availed by 75% of households, followed by 48.75%

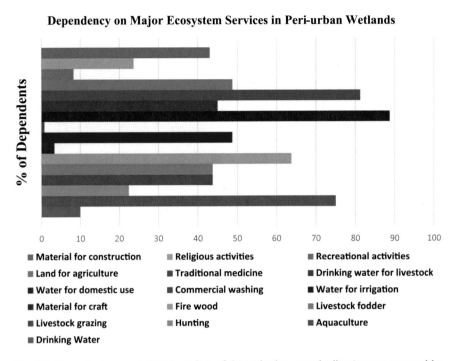

Fig. 6.5 Dependency on ecosystem services of the peri-urban waterbodies (source: prepared by authors)

of households who utilise the wetlands as agricultural space or its water for irrigation, and 43.75% who use the land for livestock grazing. The share of peri-urban households using waterbodies for domestic washing and bathing was found to be 88.75%, and that of households collecting traditional medicines and food from the wetland areas was 81.25%. Likewise, 63.75% of households were collecting firewood, and 43% were collecting construction materials from the wetland areas (Fig. 6.5).

The assessment found that gender equity in accessing the ESs of the wetlands depends on the nature of the service. Females are the predominant users of those waterbodies which are mainly used for domestic purposes and small-scale fish farming at the household level. On the other hand, farming activities in waterbodies that are used for commercial fisheries are mostly dominated by males. When asked to rank the importance of the waterbodies, 73.75% of respondents reported that the waterbodies are of prime importance in their day-to-day lives, as they support their livelihoods and fulfil basic needs as well. Again 68.75% of respondents reported that degradation of the wetlands has affected their livelihoods. Thus, the study revealed that the peri-urban wetland SEPLS are still of prime importance to the peri-urban community. They can be termed the "lifeline" of the peri-urban population.

3.3 Vulnerability and Sustainability Factors

The sociometric study, conducted pre-, post-, and during monsoon periods in 2018–2019, revealed that LVI is the highest in the urban fringes of both wetland SEPLS. Overall, in EKW it was 0.526, and in DBW it was 0.545. In EKW, vulnerability was the lowest for the food security component and also very low for the social network component. This is due to intensive culture fisheries and paddy cultivation in EKW, which causes habitat damage due to overusage of chemical fertilisers and feed, loss of local agro-biodiversity, as well as deterioration of ecosystem services. However, in Deepor Beel, only catch fisheries and livestock rearing support livelihoods. The literacy rate in households near the urban fringe is higher than literacy rates in far-off households. Thus, beneficiaries staying near urban settings have more access to societal networking and feel less vulnerable than those who are settled far from the urban areas, as they do not have at-hand access to urban facilities. For example, residing near urban settlements assures better access to water supply and health services, thereby reducing vulnerability. However, affordability is also an issue for the marginal communities to access these. Unfortunately, this has prompted a trend among marginal households to clutter near urban areas apparently for better access to crucial amenities like health services and water supply. As a spin-off effect, this has led to anthropogenic encroachments and deterioration of the wetland ecosystem services on which these marginal communities usually depend for their livelihood sustainability. Therefore, these chains of events further augment their vulnerability, though they are in close proximity to urban areas. This trend also suggests that due to overcrowding, urban fringes get more polluted and face greater water scarcity and finally access to basic amenities is diminished for the urban poor. Detailed analyses have suggested that when dependency on nature's services is high, socioeconomic vulnerability is also high in areas near to urban settlements. Vulnerability was also found to be high when heads of the household were female or illiterate, as well as for households made up of unskilled family members. Female heads of households are less aggressive in seeking out amenities or services owing to the societal dominance of males, while illiterate and unskilled heads of households were incapable of acquiring opportunities in a competitive environment. Higher dependence on agriculture, migration of male members due to increasing LUP changes, conversion of agricultural to non-agricultural land, and greater proximity to urban areas seemingly increased societal vulnerability, as in the case of DBW. Adopting more than one type of agricultural livelihood is one of the adaptive strategies for agricultural communities in these SEPLS in coping with climate vulnerability and decreasing viability of agriculture. Thus, families having multiple agricultural practices were considered less vulnerable, which is why EKW had a better LVI score than DBW. A community's livelihood vulnerability was found to be related to the degree of displacement of people from their traditional livelihoods, as in the case of EKW. Most families have at least one member in a non-agricultural livelihood, working in unorganised sectors in the urban areas. Due to the lack of reliance on agriculture, the community has adopted such livelihood strategies. Many of the farmers practice farming in leased lands and commute to these places from other parts of the peri-urban areas of both Kolkata and Guwahati. The percentage of households having no fertile land is the highest in DBW, whereas fish farms are sharply shrinking in EKW.

3.4 Bio-Rights: Impact Assessment

The rights-based neo-economic conservation model was initiated by the South Asian Forum for Environment (SAFE) in the year 2016–2017 to recognise the rights of the marginal fishers and farmers in the wetland SEPLS, who are also engaged in the conservation and maintenance of wetland habitat. To improve the wetland habitat, capacities were built in conservation activities among the wetland communities, like maintenance of water quality, regulation of nutrients leading to eutrophication, planting of wetland species for phytoremediation, stabilisation of embankments of lakes, and maintenance of the environmental flow of water. To compensate for the opportunity costs of beneficiaries, they were trained in alternative livelihood opportunities like making handicrafts out of water hyacinth plant fibre, recycling of municipal solid waste that spoils the habitat, and organic family farming.

Assessments were conducted over a 3-year span in 2016–2017 and 2018–2019. The recognition as a formal local institute and the financial inclusion of the beneficiaries entailing insurance coverage and credit linkages compensated for opportunity costs from payments for ecosystem services (PES), thereby assuring livelihood security, financial inclusion, and risk mitigation. Formation of Joint Liability Groups (JLGs) comprised of the fishers and farmers of the wetland communities was initiated for ensuring shared responsibility and a strengthened local institutional framework. Financial inclusion of these JLGs through bank linkage, credit linkage, and micro-insurance coverage spread the risk and ensured economic security as well. Gender equity and women's empowerment were core components of the intervention. Three villages each in EKW (EKW V1, V2, V3) and DBW (DBW V1, V2, V3) were selected for the intervention, comprising a total of 2500 households in the six villages. The villages were assessed on indicators like LVI, increased primary productivity, and revenues earned therefrom, as well as additional man-days created through alternative livelihood opportunities. Impacts on habitat health were accounted for using the Trophic State Index (TSI) in the conserved wetlands.

Review and analysis of data available from 2016 to 2017 and from April to December 2019 showed the striking impacts of the bio-rights intervention on both of the Ramsar wetland SEPLS (Fig. 6.6). These impacts are highlighted below:

(a) Strengthening of local institutions and enabling of risk coverage through micro-insurance and financial inclusion substantially lowered vulnerability in the 3 years of time span.

(b) Community-led conservation in both SEPLS, viz. EKW and DBW, accentuated sustainable intensification of ecosystem services, thereby increasing the returns

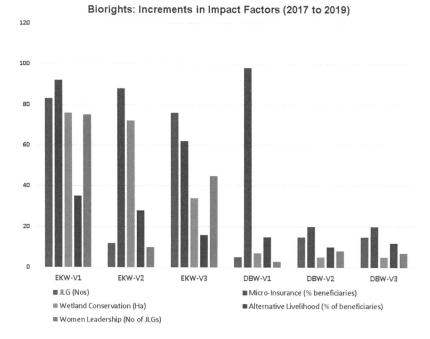

Fig. 6.6 Bio-rights: increments in impact factors (source: prepared by authors)

on revenues, as evidenced by an increase in the per capita income of fishers by 17–20% per year. The bio-rights intervention also leveraged alternative economic opportunities from ecotourism and circular economic interventions for women's entrepreneurship.

(c) Collective responsibility in wise-use practices and shared efforts for sustainable intensification of ecosystem services were more distinct in EKW, owing to the organised patterns of community engagements and robust local institutions. This is thus considered as an indicator of sustainability, accomplished through bio-rights.

(d) Conservation activities in the SEPLS improved habitat health and increased primary production in the area.

3.5 Assessment Studies During the Pandemic Crisis

During the COVID pandemic, announcement of a nationwide lockdown triggered a mass reverse exodus of temporary migrants working as casual labourers and an enormous loss of livelihood in the informal sector. During this crisis, nearly 2000 landless labourers belonging to the wetland communities returned to EKW and DBW. The bio-rights intervention could minimise the adversity both on life and

HEALTH SERVICES OF SEPLS
5-POINT ATTITUDE SCALING IN BENEFICIARIES

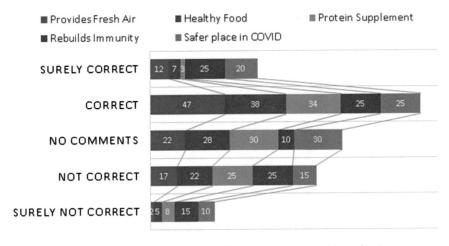

Fig. 6.7 Attitude scaling on health services of SEPLS (source: prepared by authors)

livelihood of these migrants without compromising the conservation objectives. The intervention supported these migrants with some economic opportunities through manual labour work in farming and fisheries, and engagement in the supply chains of the farming sectors. While the intervention extended health support to the urbanites living in the near vicinity by providing protein supplements in daily food from table fish, fox nuts, and duckery products, production of fresh vegetables and local medicinal plants also supported indigenous communities in the wetlands during the crisis. The agro-biodiversity preserved in the local cropping cycles, which are mostly managed by women farmers in the wetland SEPLS, is the source of food and food supplements for augmenting the health and immunity of the local people. This had a great bearing on the lives of these indigenous communities during the pandemic crisis. The urbanites in the vicinity, who visited the wetland ecotourism trails for morning walks and physical activities during "work-from-home" periods in the pandemic, experienced the ambience of fresh air, oxygenated and free from urban pollution. An attitude scaling on the health impacts of these eco-trails is indicated in Fig. 6.7, showing that 50–60% agreed that the landscape rebuilds immunity and provides fresh air, while 37% deemed the protein food supplements made available from these wetlands as important.

4   Outcomes and Impacts

Wetlands, though alienated from mainstream conservation owing to deficient policy frameworks, are unique socio-ecological production landscapes and seascapes (SEPLS) which are significant in the milieu of climate change. This chapter reviews two peri-urban Ramsar sites to put forward ecosystem-based adaptive strategies for conservation and recommends a framework using a rights-based neo-economic approach to compensate for the opportunity costs of wetland inhabitants and dependents. This chapter makes a case for this approach based on the following observations:

(a) It is observed that the vulnerable communities in these SEPLS continually explore opportunities amidst challenges imposed on them, such as during the pandemic when they attempted to reduce the impact of externalities through appropriate resource stocks, extraction, and marketing strategies. Since most of the beneficiaries work in informal sectors, wherein institutional arrangements are somehow fragile and cannot handle onslaughts imposed by large-scale and systemic perturbations, beneficiaries chose either to accept working as casual labourers in urban areas or to change to alternative economic opportunities in urban service deliveries, as evidenced from sociometric assessments in the wetland communities.

(b) Perusal of outcomes from focus group discussions and household surveys in the SEPLS revealed that under economic pressure, responsibilities of risk mitigation move from individual to collective action, enabling the sharing of economic losses by suitably changing livelihood options. However, mechanisms to monitor compliance and enforcement of the norms for collective community action are not easily practised due to a diverse perception of risks. Obviously compensation for loss and damage comes mostly from the use of nature's goods and services. This leads to loss of habitat and biodiversity owing to overuse.

(c) Predictive analyses based on participatory evaluation of ecosystem services and the fate of wetland functions under worst-case scenarios can also greatly help to define best scenarios in bio-rights-centred livelihood promotion coupled with community conservation strategies and related institutional mechanisms developed in the course of implementation of the bio-rights project.

(d) By tracing the biodiversity, community dynamics, and relevance of spatio-temporal changes in LUP and LC, as well as institutional arrangements that have formalised with the management of the EKW and DBW over years of community-based conservations in these two SEPLS, the chapter aspires to explore avenues to design a more inclusive sustainable ecosystem service-based management plan that includes novel concepts like bio-rights.

5 Recommendations for a Framework

Reviewing the determinants of the dependence on wetland ecosystem services by marginal agrarian communities vis-à-vis the impacts of urban encroachment on these peri-urban wetlands in advocating sustainable intensification of primary productivity, a rights-based conservation framework has been recommended for wetlands near urban growth centres for adaptive policy planning (Amerasinghe & Dey, 2018). A brief overview of this framework is outlined hereunder:

1. In recognising the importance of these peri-urban wetland habitats and their services, a normative practice has to be enunciated to measure the economic value of goods and services, ensuring that the estimations are non-ambiguous and non-discriminatory at all levels of policy and practice to leverage the benefits of conserving these wetlands, especially in the climate milieu.
2. The peri-urban wetlands must be recognised by town planners and urban architects as socio-ecological production landscapes, and the returns on investments in intensifying primary productivity in these wetlands must be added to the GDP. This would enable sharing of benefits, and co-benefits earned from the goods and services of the wetlands with the marginal communities to ensure inclusive growth.
3. Ecosystem-based adaptation for conservation of peri-urban wetlands needs to be included as mandatory clauses in town planning and landscaping, leaving no one behind in participatory management, and be recommended for enriching national natural capital like forests and rivers.
4. The lessons learned must be incorporated in the draft for developing National Wetland Policy (conservation and management) in India for inclusion and adoption.

The following facets may be considered for further reference to be included in the draft as well as the knowledge economy. Reference frame for defining community-based wetland conservation strategies and enabling a platform for redressal of issues pertaining to it:

(a) Wetland Conservation

1. Conservation strategies for peri-urban wetlands are to be outlined at the community-ecosystem interface to recognise community governance in conservation, similar to policies and practice utilised in community forest management.
2. The local wetland authority entrusted for the conservation and management of wetland SEPLS has to be inclusive and should have equitable community representation for redressal of issues as well as assure remedial support to ensure equity, reciprocity, and partnership for conservation.
3. A frame of reference is expected to be in place to assure conservation and wise use of peri-urban wetlands through a national policy document, to be referred to as the National Wetland Policy [NWP] to mitigate any or all issues pertaining to conservation and management of these wetlands.

(b) Ownership and commons' property rights

1. With regard to the ownership and property rights of wetland ecosystems, it is recommended that they be private, joint, and/or state owned. Alterations to conservation regulations, changes pertaining to land-use pattern of the SEPLS, or changes in usages would mandatorily need to prioritise commons' bio-rights and conservation of biodiversity to upkeep the ecosystem services for all stakeholders, ensuring equitable access and benefit sharing.
2. In cognisance of the above mandates, ecosystem services of these SEPLS must be recognised as commons' property rights for economic co-benefits,

irrespective of the ownership of the wetlands, whereas efforts for conservation would be recognised as an "equal and reciprocal responsibility" for all facilitating partnership and participation beyond the property rights.

3. The conservation and management regulations and the commons' property rights of wetlands shall not alter with any alteration and/or transfer of ownership rights.

(c) Participatory conservation paradigm and bio-rights of commons

1. Planning procedures for conservation must encourage and facilitate a participatory practice, leaving no one behind to promote inclusive partnership.
2. Implementation of the conservation plan must have a robust monitoring and periodic evaluation (M&E) system and feedback systems for adaptive learning and constructive review of the plan being implemented.
3. In enforcing the implementation plan and considering wise use as conservation priority, the opportunity costs of commons incurred in forgoing the co-benefits of ecosystem services need to be estimated and compensated through novel and uniform financial models. The compensation may be used for either risk spreading or coverage, as may be decided collectively by the participants from time to time.

(d) National Wetland Policy for conservation and management

1. The state must implicate on an emergency basis the propounding of a national reference agenda for conservation and management of wetlands, to be recognised as the "National Wetland Policy" that may consider the details, as discussed above, to be integral to its operational framework.
2. The state must appoint efficient, trained, and empathetic experts as custodians of the policy upon implementation to facilitate the conservation objectives and augment ecosystem services thereto.
3. The policy framework, thus built, must seek concurrence with all national policies for conserving other potential ecosystems to undermine the conflicts of interests and opinions in overlapping jurisprudence and co-parallel regulations, so as to avoid abundance and loss of national natural resources.

Acknowledgements The authors sincerely acknowledge the support extended by the International Water Management Institute, Sri Lanka, and the Asia Pacific Network for Global Change Research, Japan, for carrying out the study.

References

Amerasinghe, P., & Dey, D. (2018). Recommendations for the wise use of urban and peri-urban wetlands in Kolkata, India. *WLE Briefing Series*, no. 23, viewed 17 October 2018. Retrieved from https://cgspace.cgiar.org/rest/rest/bitstreams/4bd3cd70-de73-49d0-9909-25580880a4af/ retrieve.

American Public Health Association (APHA). (1995). *Standard methods for the examination of water and wastewater* (19th ed.). American Public Health Association.

Bansal, S. (2020). Mumbai lost 71 percent of wetlands in last four decades: report. *India Water Portal*, viewed on 7 March 2020. Retrieved from https://www.indiawaterportal.org/articles/mumbai-lost-71-percent-wetlands-last-four-decades-report.

Bassi, N., Kumar, M. D., Sharma, A., & Pardha-Saradhi, P. (2014). Status of wetlands in India: A review of extent, ecosystem benefits, threats and management strategies. *Journal of Hydrology: Regional Studies, 2*, 1–19. https://doi.org/10.1016/j.ejrh.2014.07.001

Bhattacharyya, A., Sen, S., Roy, P. K., & Mazumdar, A. (2012). A critical study on status of east Kolkata wetlands on special emphasis on water birds as bio-indicators. In M. Sengupta, & R. Dalwani (eds.), *The 12th World Lake Conference (Taal) Proceedings*, 28 October-2 November, Jaipur, India, pp. 1561–1570.

Carlson, R. E. (1977). A trophic state index for lakes. *Limnology and Oceanography, 22*(2), 361–369. https://doi.org/10.4319/lo.1977.22.2.0361

Davies, S. P., & Tsomides, L. (2014). *Methods for biological sampling and analysis of Maine's Rivers and streams*. Maine Department of Environmental Protection Bureau of Land and Water Quality, Division of Environmental Assessment.

de Groot, R., Fisher, B., Christie, M., Aronson, J., Braat, L., Haines-Young, R. Y., Gowdy, J. M., Maltby, E., Neuville, A., Polasky, S., Portela, R., & Ring, I. (2010). Integrating the ecological and economic dimensions in biodiversity and ecosystem service valuation. In P. Kumar (Ed.), *The economics of ecosystems and biodiversity: the ecological and economic foundations* (pp. 9–40). Earthscan. http://teebweb.org/wp-content/uploads/2013/04/D0-Chapter-1-Integrating-the-ecological-and-economic-dimensions-in-biodiversity-and-ecosystem-service-valuation.pdf

Dey, D. (2008). 'Biorights' of commons as an economic opportunity for negating negative link between poverty and nature degradation', Digital Library of Commons, Indiana University, viewed on 12 September 2008. Retrieved from https://dlc.dlib.indiana.edu/dlc/bitstream/handle/10535/1874/Dey_124901.pdf?sequence=1.

Dey, D., & Amerasinghe, P. H. (2022). Community "bio-rights" in augmenting health and climate resilience of a socio-ecological production landscape in peri-urban Ramsar wetlands. In M. Nishi, S. M. Subramanian, & H. Gupta (Eds.), *Biodiversity-health-sustainability nexus in socio-ecological production landscapes and seascapes (SEPLS)* (pp. 107–127). Springer Nature.

Ghosh, A. K. (1990). Biological resources of wetlands of east Kolkata. *Indian Journal of Landscape System and Ecological Studies, 13*, 10–23.

Ghosh, D. (2005). *Ecology and traditional wetland practice: Lessons from wastewater utilisation in the East Calcutta wetlands*. Worldview.

Ghosh, D., & Sen, S. (1987). Ecological history of Calcutta's wetland conservation. *Environmental Conservation, 14*(3), 219–226.

Ghosh, S., & Das, A. (2020). Wetland conversion risk assessment of East Kolkata wetland: A Ramsar site using random forest and support vector machine model. *Journal of Cleaner Production, 275*, 123475. https://doi.org/10.1016/j.jclepro.2020.123475

Hahn, M., Riederer, A., & Foster, S. O. (2009). The livelihood vulnerability index: A pragmatic approach to assessing risks from climate variability and change—A case study in Mozambique. *Global Environmental Change, 19*(1), 74–88. https://doi.org/10.1016/j.gloenvcha.2008.11.002

Mahadevia, D., Mishra, A., & Joseph, Y. (2017). Ecology vs housing and the land rights movement in Guwahati. *Economic & Political Weekly, 52*, no. 7.

Mcinnes, R. (2014). Recognising wetland ecosystem services within urban case studies. *Marine and Freshwater Research, 65*(7), 575. https://doi.org/10.1071/MF13006

Mitra, S., & Bezbaruah, A. N. (2014). Railroad impacts on wetland habitat: GIS and modeling approach. *Journal of Transport and Land Use, 7*(1), 15–28. https://doi.org/10.5198/jtlu.v7i1.181

Mukherjee, P. (2011). Stress of urban pollution on largest natural wetland ecosystem in East Kolkata-causes, consequences and improvement. *Archives of Applied Science Research, 3*(6),

443–461. http://scholarsresearchlibrary.com/archive.html) ISSN 0975-508X CODEN (USA) AASRC9.

Osinuga, O., & Oyegoke, C. O. (2019). Degradation assessment of wetlands under different uses: Implications on soil quality and productivity. *African Journal of Agricultural Research, 14*(1), 10–17.

Reddy, S. C., Saranya, K., Shaik, V. P., Satish, K. V., Jha, C. S., Diwakar, P. G., Dadhwal, V. K., Rao, P. V. N., & Murthy, Y. V. N. K. (2018). Assessment and monitoring of deforestation and forest fragmentation in South Asia since the 1930s. *Global and Planetary Change, 161*, 132–148. issn:0921-8181. https://doi.org/10.1016/j.gloplacha.2017.10.007

Saikia, J. L. (2019). Deepor Beel wetland: Threats to ecosystem services, their importance to dependent communities and possible management measures. *Natural Resources and Conservation, 7*(2), 9–24. https://doi.org/10.13189/nrc.2019.070201

U.S. Geological Survey (USGS). (2006). *National field manual for the collection of water-quality-data: U.S. Geological Survey Techniques of Water-Resources Investigations*, Book 9, Chapters A1–A9. Retrieved from https://pubs.water.usgs.gov/twri9A.

The opinions expressed in this chapter are those of the author(s) and do not necessarily reflect the views of UNU-IAS, its Board of Directors, or the countries they represent.

Open Access This chapter is licenced under the terms of the Creative Commons Attribution 3.0 IGO Licence (http://creativecommons.org/licenses/by/3.0/igo/), which permits use, sharing, adaptation, distribution and reproduction in any medium or format, as long as you give appropriate credit to UNU-IAS, provide a link to the Creative Commons licence and indicate if changes were made.

The use of the UNU-IAS name and logo, shall be subject to a separate written licence agreement between UNU-IAS and the user and is not authorised as part of this CC BY 3.0 IGO licence. Note that the link provided above includes additional terms and conditions of the licence.

The images or other third party material in this chapter are included in the chapter's Creative Commons licence, unless indicated otherwise in a credit line to the material. If material is not included in the chapter's Creative Commons licence and your intended use is not permitted by statutory regulation or exceeds the permitted use, you will need to obtain permission directly from the copyright holder.

Chapter 7
Effective Water Management for Landscape Management in the Siem Reap Catchment, Cambodia

Chris Jacobson, Jady Smith, Socheath Sou, Christian Nielsen, and Peou Hang

Abstract International awareness of the world-renowned Angkor Wat temple complex has drawn attention to the challenges of climate change, deforestation, and water management in Cambodia. The aim of this chapter is to examine the benefits of enhanced water management provided within Angkor Archaeological Park, and to consider challenges to maintaining those benefits. The Authority for the Protection of the Site and Management of the Region of Angkor (APSARA) designed and in 2014–2018 implemented within the Park a water management project to recharge groundwater supplies, mitigate floods, and provide irrigation. To assess the benefits, we draw on an economic analysis of ecosystem service changes, including interviewing 145 households from across the Park and four experts. To assess challenges to sustaining the benefits, we also interviewed 73 households and conducted 12 focus group discussions in the upper catchment. We used a combination of quantitative analysis (i.e. economic assessment) and qualitative data analysis (e.g. thematic analysis). Our analyses of data from people living in and around the Park showed that improved water management (e.g. reinforcing dykes and storage facilities for groundwater recharge) and investment in economic diversification (e.g. tourism, horticulture, and heritage crops projects) reduced vulnerability of the people to climate hazards. Currently, these benefits are threatened by forest loss in

C. Jacobson (✉)
Live and Learn Environmental Education, Melbourne, VIC, Australia

The University of Queensland, St. Lucia, QLD, Australia
e-mail: chris.jacobson@livelearn.org

J. Smith · C. Nielsen
Live and Learn Environmental Education, Melbourne, VIC, Australia

S. Sou
Live and Learn Environmental Education, Phnom Penh, Cambodia

P. Hang
Authority for the Preservation and Management of Angkor and the Region—APSARA, Siem Reap, Cambodia

© The Author(s) 2022
M. Nishi et al. (eds.), *Biodiversity-Health-Sustainability Nexus in Socio-Ecological Production Landscapes and Seascapes (SEPLS)*, Satoyama Initiative Thematic Review, https://doi.org/10.1007/978-981-16-9893-4_7

the upper Siem Reap catchment. Our analysis of data from the upper catchment showed that forest loss also resulted in detrimental effects to human health and well-being, and was associated with higher food insecurity. Solutions are suggested to enhance nature-based employment opportunities and promote economic diversification. This would extend the scope of management of this socio-ecological production landscape (SEPL) beyond the Park and ensure its sustainability by improving the health and well-being of the people living in the upper catchment.

Keywords Angkor, Siem Reap · Diversification · Climate change · Food security · Ecosystem services, Nature-based solutions

1   Introduction

Management of human-nature interactions has occurred across millennia in the Angkorian landscape (Siem Reap Province). Angkor Archaeological Park is one of the most important archaeological sites in Southeast Asia. It contains the remains of various capitals of the Khmer Empire. In 1992, Angkor Archaeological Park was recognised by the United Nations Educational, Scientific and Cultural Organization (UNESCO) as a world heritage site. Its listing highlights not only archaeological heritage, but also the linkages between tangible heritage (e.g. temples) and intangible heritage (e.g. forest conservation, water management). Thus, the Park is a socio-ecological production landscape (SEPL) based within one of the largest archaeological sites in the world.

The Park is threatened by (1) limited protection of upstream ecosystems, which impacts the groundwater recharge and availability within the Park, and (2) pressure from communities in the surrounding rural areas, for whom harvesting natural resources (such as forests) is an easy but unsustainable form of livelihood. In the upper Siem Reap River catchment, forest degradation resulted in loss of 36.6% of forest area between 1989 and 2019, and changes in structure and composition (Chim et al., 2019; Wales, 2020). Most of the loss resulted from conversion for agriculture (Chim et al., 2019) due to increased population pressure on the land, as evident in decreased landholding size (National Institute of Statistics, 2009, 2017). These changes have increased run off, increasing sediment levels in the lower catchment ponds and lakes, with potential impacts on fishery productivity (Chim et al., 2021). The result is an overall trend towards a drier landscape with less precipitation (Jacobson et al., 2019) and changes to the flood pulse in the Tonle Sap lake (Frappart et al., 2018). In addition, there is increased water consumption for household use and agriculture, which lowers groundwater levels and consequently reduces supply, and potentially temple stability (Kirsch, 2010; Chim et al., 2021).

In response to these challenges (i.e. agricultural expansion, deforestation, and changes in water availability), the Ministry of Foreign Affairs and Trade of New Zealand (MFAT) supported the Authority for the Protection of the Site and Management of the Region of Angkor (APSARA) to conserve and restore an archaeological hydraulic engineering system and support livelihood advancement projects within the Park (2014–2018). However, the benefits of these activities have as yet not been documented. In addition, the importance of application of effective management systems beyond the Park boundaries has not been assessed. This chapter examines the potential effectiveness of the project's activities to enhance the resilience of this SEPL. It demonstrates how project interventions addressed the nexus between ecosystem health and human well-being, and the importance of their extension beyond the Park boundary. The case is significant as a study of reinvigoration of traditional water and landscape management, demonstrating the benefits from combining modern engineering technologies with traditional knowledge through heritage conservation.

2   Background and Methods

2.1   The Angkor World Heritage Site and Its Significance

Situated in the province of Siem Reap, Angkor Archaeological Park, "the Park" (Fig. 7.1 and Table 7.1), is one of the most important archaeological sites in Southeast Asia, as well as a major tourist attraction (Hang et al., 2016). The Park stretches over 400 km^2 and contains the remains of various capitals of the Khmer Empire which flourished over six centuries from the early ninth century. It includes temples (e.g. world-renowned Angkor Wat), hydraulic structures (e.g. basins, dykes, reservoirs, canals), and forested areas. Angkor is considered a "hydraulic city" because of its complicated water management network used for systematically stabilising, storing, and dispersing water throughout the area. The temple complex is supported by three river systems stretching 50 km upstream. Lakes, moats, ponds, and royal basins saturate the sand-clay soils supporting the temples, ensure temple stability, enable irrigation, and support the food security of the increasing local population; they also provide flood protection to Siem Reap city (Hang et al., 2016; Chim et al., 2019). The Angkorian water management system includes a series of moats and lakes with spillways, connected by canals (see Chim et al., 2021 for a diagrammatical representation). Water gates direct and control water flows, while intact canals, reservoirs, and spillways direct water to or away from where it is needed for agriculture purposes. Canals whose walls are partially eroded result in a loss of water from those canals. Moats and lakes recharge groundwater, and provide water storage, limiting floods in the wet season and mitigating against drought in the dry season. Recent LiDAR analyses (e.g. Chen et al., 2017; Evans, 2016; Wales, 2020) provide scientific analysis of the spatial scale of these structures beyond the Park. The analysis demonstrated that ancient water management occurred across the whole landscape, not just within the Park boundaries as currently occurs.

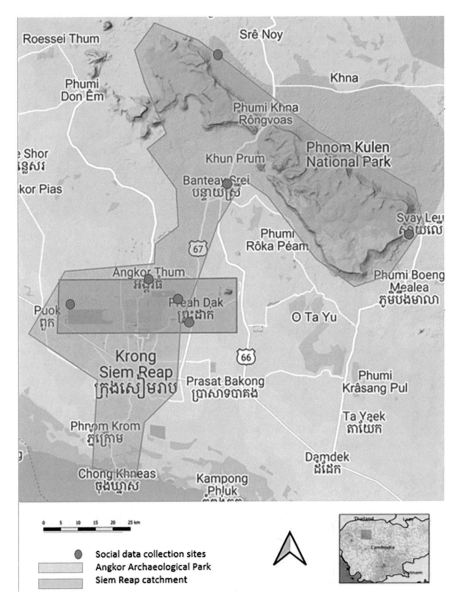

Fig. 7.1 Map of the Angkor Landscape showing the core zone of the Park (approximately 210 km^2) and the catchment (approximately 1220 km^2) (polygon developed by authors, source: Map data © Google, 2021; insert from Nakau, 2020)

Table 7.1 Basic information of the study area

Country	Cambodia
Province	Siem Reap
District	Project site: Angkor Thom, Siem Reap, Prasat Bakong, Banteay Srey Catchment: Angkor Thom, Banteay Srey, Siem Reap, Svay Leu, Varin, Prasat Bakong
Municipality	n.a.
Size of geographical area (hectare)	Site: 6900 Catchment: 122,300
Number of direct beneficiaries (persons)	Park: 118,652
Number of indirect beneficiaries (persons)	1,014,000 (2019 census)
Dominant ethnicity(ies), if appropriate	n.a.
Size of the case study/project area (hectare)	21,000
Geographic coordinates (latitude, longitude)	13°21′47.9″ N, 103°51′23.0″ E

2.2   Angkor Park Management Projects

In partnership with APSARA, the New Zealand MFAT has invested in making the management of the Park more resilient since 2010. This has included development of a Park plan and enhancement of capacity for community-based natural resource management within the Park. In 2014, the Angkor Community Heritage and Economic Advancement Project (ACHA) (2014–2019) was initiated. The goal of this project was to increase food security and promote sustainable management, including revitalisation of ancient water management across the entire Park. An evaluation of this project provided understanding of the benefits of the water management component. The ACHA project activities included (1) 16 new or repaired gates, and rehabilitation of dykes, moats, and spillways, and (2) enhanced water storage through the new and rehabilitated storage structures ($>$1,700,000 m^3 storage) (Fig. 7.2 provides an example). Water storage activities were designed to ensure soil water saturation, support temple stability, as well as capture water for irrigation during the wet season ($>$1730 ha irrigated paddy under cultivation) and in the dry season (20 ha irrigated paddy under irrigation). This chapter documents the benefits of these water management activities to the SEPL.

Fig. 7.2 Water infrastructure around the temple complex (photo by C Jacobson)

2.3 Assessing the Impacts of Existing Management and Future Opportunities

This case study was based on an assessment of ecosystem service benefits associated with the water management activities within the Park by comparing to areas beyond it. It draws on two sets of previously unpublished data.

Data Collection

We used two data sets to assess the impacts of existing management and future opportunities. The first data set was used to assess the economic benefits of ACHA water management to people living in the Park, where provisioning, regulating, and cultural services provided by the ecosystem were identified.[1] Indicators used were:

---

[1] This work was led by Jacobson et al. (2018) to provide insights for the donor, and has remained unpublished.

1. Changes in the economic value of local livelihoods—provisioning services (household interview data)
2. Changes in the value of tourism—provisioning services (calculation using the Ministry of Tourism data and data from APSARA staff interviews)
3. The value of improved flood management—regulating services (calculation using household interview data, and data on flood impact costs from ADB (2012))
4. Changes in the value of cultural services provided (calculated using transfer values)

Household interviews were used to inform the assessment of changes in local livelihoods. The Cambodian local government structure formed the basis of household interviewee selection, and is organised in district-commune-village-household hierarchy. We conducted 145 household interviews. Household selection was based on the following steps:

1. Selection of one target commune from each of the five districts, ensuring representation from the north, west, east, centre, and south of the Park where soil quality and tree density differ, also reflecting agriculture and livelihood diversity observed during field visits
2. Within the boundaries of a target commune, random selection of at least 1.5% of all households (minimum sample size of 15) stratified by villages wholly within the Park. This provides for a margin of error of $<10\%$; the ability to provide precise estimates is limited by a lack of sample frame data

These interviews were conducted with 118 participants in 2018 in Khmer language by APSARA staff. Structured questions (quantitatively analysed using descriptive statistics) were included about:

• Flood frequency (reflecting on frequency before and after ACHA water infrastructural development between 2016 and 2018) and flood impacts (before and after infrastructural development, using an ordinal scale[2])
• Livelihood activities (before and after water infrastructural development), including farmgate value (i.e. value when purchased from the farmer) or wholesale produce value (i.e. value when purchased from a wholesaler) and enterprise value (i.e. for non-agricultural produce)
• Food security, using Food Insecurity Experience Scale (Carletto et al., 2013)
• Debt

Lastly, APSARA staff were interviewed to understand the reduction in flooding frequency and extent in the Park that resulted from improved water management; this provided data on the number of days of Park closure that enabled assessment of benefits of water management associated with tourism. The four interviewed

---

[2]This research used a development project as the case study. The project lacked ex post data on flood impacts, and hence used an ordinal change analysis. The ordinal scale acknowledges the limitations in data quality that would exist had we used unverified recall data.

APSARA staff members were not involved in the household interviews on changes in flood frequency and their impacts on food security and livelihoods.

The second data set, collected in 2020,[3] provides reference values that enable comparative analysis with data on Park management. It provides understanding about broader landscape management constraints that have impact upon the Park, and the ability to maintain the benefits derived from the ACHA project. It included 73 semi-structured interviews conducted in four communes in the upper and mid-Siem Reap catchment outside of the Park (40 females, 33 males). It also included 12 focus group discussions, which engaged an additional 93 community participants (50 females, 43 males) and local government officials (3 female, 23 male) associated with each commune. This sample size was considered appropriate given the exploratory intent and qualitative nature of this research.

For the second data set, interviews were based on questions about the following topics that were qualitatively analysed:

- Current uses of natural resources
- Changes in the frequency and impacts of weather events
- Challenges to the management of land, water, and forests
- Perceptions about environmental quality, changes in it, and predictions about the future

For the second data set, focus group discussions were based on the topics of:

- Environmental quality, including changes in ecosystem services
- Livelihood benefits from native biodiversity
- Adaptation and change in livelihoods in response to shocks, changes, and/or disasters
- Engagement in decision-making and information sharing

Data Analysis Methods

For the first data set, we provide analyses of household data. We then combined this primary data with values from other Cambodia studies (ADB, 2012) to conduct economic valuation of the different types of ecosystem services affected by water management.

Average household livelihood changes were scaled to the total number of households in the Park, and then assigned an attribution range[4] of 10–20%. This attribution range was based on an expert assessment by hydrologists in APSARA, and recognised that water was one factor contributing to improved agriculture livelihoods.

---

[3]This data is presented in Jacobson and Smith (2020), and has served as an internal document shared with the donor and project partner, and remains unpublished in academic form.

[4]Attribution range refers to expert judgements (made by APSARA experts in livelihoods, water and natural resources) about the share of change that could be attributed to the ACHA project.

The value of tourism changes was calculated using ticket value, visitor numbers, and changes in Park access. Park access changes when flooding occurs and tourists are unable to visit, resulting in a loss of revenue. We asked experts to estimate the total average of days of closure before and after water management, and their responses provided conservative and upper bounds of the value of tourism changes, which we provided to represent the lack of certainty on the impacts to tourism across the entire park (i.e. a closure rate of between 2 and 4 days per year).

The value of flood protection to Siem Reap province was calculated using avoided cost estimates, based on the costs of flood impacts associated with typhoon Ketsana (ADB, 2012). This typhoon impacted Siem Reap Province before ACHA project activities occurred, but only impacted the population within the Park and downstream of it. The values from this ADB study were multiplied by the proportion of the population living in the province at the time that was impacted, and multiplied by changes in annual flood probability (expert assessments of 10% annual likelihood pre-ACHA, reduced to 1–2% after ACHA activities).

Lastly, cultural service values attributable to ACHA water management were assessed by determining existence values. We used transfer values based on the willingness to pay for cultural heritage (temples) presented in studies from Thailand and Vietnam (Seenprachawong, 2006; Tuan et al., 2009), adjusting for per capital GDP inflation since they were conducted. The per capita existence value was scaled to the entire population of Cambodia and ten million international visitors, per year. Attribution values of 5–10% account for the potential share of water management in the temples' existence.

The second data set was analysed using thematic coding, and data are presented on the themes of forest uses, water and climate, health impacts associated with landscape change, connections between upper catchment and the Park, and prospects for the future.

Limitations

The project used ex post techniques[5] to evaluate the impacts of a development project. One limitation of this technique is the breadth of indicators that can be included. While additional indicators and specific measures (e.g. health and biodiversity metrics) are relevant and potentially of interest to many involved in SEPL management, a lack of reference data for comparing change makes them less meaningful for impact assessment. The use of data from the upper catchment is primarily to contextualise the impact of the project, and its sustainability. The

---

[5] Ex post is a form of evaluation typically used to justify whether a specific intervention has worked, its strengths, and its weaknesses. It is conducted after a project has been completed. Ex post uses change measures for key indicators. However, not all potential project benefits or impacts of a project can be foreseen when a project begins. In these cases, ex post techniques compare achievements to reference areas. JICA (2004) notes that ex post evaluation provides for impact and sustainability evaluation.

limitations of use of data from the upper catchment to assess benefits from a project implemented in the Park are acknowledged.

While health was not an explicit focus during data collection, the qualitative analysis does highlight the linkages between water and health, focusing on health in terms of the well-being of communities. Likewise, we use the forests and their protection (or lack thereof) as an indicator of biodiversity.

3   Results

3.1   Understanding the Impacts of Water Management in Angkor Park

Interviewees' responses indicated a reduction in flood frequency and impacts following implementation of the ACHA project. Compared to 2013 before the infrastructure investment, 49.6% of survey respondents reported decreased flood frequency, compared to 21.0% who reported increased frequency. Meanwhile, 47% of the respondents observed reduced impacts and 39.1% completely averted floods, whereas 25.2% reported increased impacts. Of these, 5.2% reported experiencing a flood where they had not seen one previously.

An analysis of interview data revealed that between 2014 and 2017, average household income also increased, driven primarily by agricultural development. Agriculture-related increase was primarily due to increase in vegetable production, rising from an average of 190 USD per household in 2014 to 1592 USD in 2017. Despite the increased income, 32.3% of households had a member experiencing severe food insecurity, and 59.4% of households had debt. Reference values for food insecurity in the same year from the upper catchment as obtained from Jacobson et al. (2019) and Jacobson and Seng (2018) unpublished data are higher, being 51.7% and 72.0%, respectively, for the two studies.

Economic analysis methods provide total estimates of the economic value arising from water management, for the life of the ACHA project, in USD calculated by Jacobson et al. (2018). These (in USD) were[6]:

- Provisioning services—local livelihoods $1,117,258–$2,234,517
- Provisioning services—tourism $3,240,000–$3,780,000
- Regulating services—flood protection $3,292,800–$3,704,000
- Cultural service benefits $1,850,000–$3,700,000

Improvements in water management through infrastructure and ancient water management reinvigoration are therefore estimated to have had immediate impacts of 9,500,308–13,429,417 USD over the project time horizon. Reduction in flood

---

[6]The range accounts for the expert judgement in attribution to ACHA project, as described in Sect. 7.2.3 on methods.

Fig. 7.3 Mild flooding in Banteay Srey, prior to water infrastructure refurbishment (photo by C Jacobson)

frequency and impact severity accounted for a third of all economic benefits. Figure 7.3 provides an example of minor flooding before the project. These benefits are significant to the province's inhabitants given the low per capita GDP. However, the ability to sustain these levels of benefit over the long term depends on maintaining water flows and water management systems.

3.2   Understanding the Broader Socio-Economic Production Landscape

The second data set (which includes household interviews and focus groups, see Sect. 7.2.3.1) enables comparative evaluation about how upstream landscape management impacts the ecosystem service delivery within the Park, and the ability to maintain the benefits derived from the ACHA project.

Forest Uses

Focus group discussions identified ecosystem services provided by forests, including:

- Attracting water[7] (humidity for improved cassava and cashew nut yield, and groundwater recharge)
- Reducing erosion and preventing sedimentation in waterways
- Limiting the speed of water movement across the landscape, and hence reducing the likelihood of flooding
- Providing livelihood benefits, including non-timber forest product collection and tourism
- Increasing soil fertility, contributing to organic fertiliser, and reducing the need for synthetic fertiliser
- Limiting the potential impacts of storms
- Providing health-related benefits including shade for heat relief, and reducing dust
- Offering shelter for wildlife
- Providing medicinal plants and spiritual trees

Forest cover in the upper catchment was observed to have declined over time due to encroachment and illegal forest uses, forest fires, and some overharvesting of natural resources (e.g. rattan), affecting the services forests provide. Key threats identified during focus group discussions included deforestation, illegal activities in protected forests, ineffective management (e.g. access to water resources), limited engagement of communities in forest management, overuse of agricultural inputs, and illiteracy of community members (i.e. affecting education and the adoption of more sustainable livelihoods).

Water and Climate

Interviews revealed that community members perceived water resources to be a critically important part of the landscape, linked to forest health. Focus group discussions revealed that the river, natural springs, and rain-fed ponds provided benefits, including:

- Water for household, crop, and livestock use
- Humidity for improved agricultural yield

---

[7]Based on the authors' experiences, many community members without scientific training, from both developed and developing countries, express the belief that forests "attract" water. In Cambodia, this could be due to a range of reasons. Alternatives to forests such as intensive agriculture might have a higher use of available groundwater than forests (which tend to have deeper root systems). It might also be due to the implications of humidity being more apparent without shade. We choose to keep the language and meaning of participants.

- Fish habitat
- Temple stability and culturally important sites[8]
- Tourism attractions (e.g. waterfalls)
- Temperature reduction

During focus group discussions, participants commented on changes in the timing, length, and intensity of the wet and dry seasons. In the upper catchment, forest degradation was identified as a driver of change in the area becoming drier with less ground and stored water availability. Participants reported that the quality and quantity of water resources had declined over time. Spring-fed water resources and wells reportedly became shallower, in some cases completely drying out. Water infrastructure was also apparently ageing and becoming less efficient, with impacts on agricultural yields and crops (e.g. the ability to grow vegetables in the dry season). Agricultural intensification, including increased pesticide and herbicide use, was associated with impacts on aquaculture, such as a decline in fish and crabs in rice fields. Increased variability in weather also reportedly resulted in chicken deaths. Successive drought and heavy rain were identified as exacerbating the risks of flooding due to soils becoming hydrophobic, and were identified as resulting in increased microfinance debt for some households given lower yields associated with these conditions. The following quote is an example from these discussions:

> Since last year, degradation of forest in the Park has caused surrounding areas to become drier. This affected yield of cashew nuts and water availability of *Teuk chub* [spring water]. (Men's focus group, Svay Leu District)

Health Impacts Associated with Landscape Change

During focus group discussions, participants identified health benefits as an ecosystem service provided by forests. Changes in forest extent were identified as increasing heat stress in both humans and animals, and loss of trees from roadsides and around houses and villages was identified as resulting in increased dust, affecting respiratory health. Participants associated the increased use of chemical fertilisers and pesticides to improve crop yield with pollution of limited freshwater resources (including spring-fed ponds, lakes, and wells). Increased frequency and intensity of storms were also associated with health impacts, including waterborne illnesses. The following quotes exemplify these discussions:

> Before the water sources were clean, but now it is so bad and that is why we need to use water filters to clean the water for household consumption. (Women's focus group, Svay Leu District)
>
> The agricultural waste and chemical use flowed into the streams and badly affected the villagers' health, like diarrhoea, itch, etc., when they used that polluted water. (Men's focus group, Varin District)

---

[8] Additional temples exist in the upper catchment outside of the Park boundary.

Connections Between Upper Catchment and the Park

Participants identified linkages between the upper and lower parts of the catchment (where the Park is situated). Focus group discussions included comments about nutrient flows linked to soil nutrients, groundwater quality, and agricultural input use. The use of "traditional" compost was identified as an option to mitigate the impacts of agricultural inputs on water, and maintain soil nutrients. Some participants were aware of the connection between their interaction with the landscape and its effect on lower parts of the catchment (i.e. the Park). One example is provided below:

> [Soil erosion] could flow to fill in the lake from year to year, and the water level would become lower and lower and cause water shortage in the next year. And the poor families would lose some foods (fish) from this bad event. (men's focus group, Prasat Bakong District)

Prospects for the Future

Interviews revealed that community members have already adapted livelihoods from a predominant reliance on rice to other crops, including cassava and cashew nut. Some participants had also adapted from growing cassava and cashew to growing other crops (e.g. vegetables, fruits) because of low yield. For example, some participants replaced cashew with orange trees (high value) and other participants had responded to soil quality depletion associated with cassava by rotating crops, and growing other cash crops such as corn and watermelon. Livelihoods have also been diversified to include chicken, duck, and pig raising.

Inequitable sharing of benefits from both forests and water was identified during interviews. For example, changes in forests were often attributed to "outsiders" who collected non-timber forest products and burnt the forest (for unclear purposes). Focus group discussions emphasised the limited ability of government or community members to enforce Community Forestry Area regulations, affecting the ability of local communities to maximise the potential benefits of forest protection. Community Forestry Areas and Community Protected Areas are established under the Forest Law and sub-decree on Community Forest Management, and the Law on Natural Protected Areas, respectively. These laws enable community committees to request areas to be recognised as such, and to recommend regulations to the government for endorsement, and subsequently enforcement. This typically includes the sustainable use of areas, including subsistence use of non-timber forest products, and use of areas for cultural purposes, in accordance with community plans. Twenty-one Community Forestry Areas and five Community Protected Areas exist within the catchment.

Over half of interview participants (58.7%) reported that natural resources were only sometimes sustainably managed. Less than half (39.1%) expected environmental quality to stay the same or improve in the future. During focus groups, participants identified the need for community-based natural resource management capacity development, better enforcement of forest regulations, further investment in water infrastructure, and governance transparency for long-term sustainability. The following quote captures these ideas:

> Community forest was protected because the villagers had understood the benefits of forest that could provide non-timber forest products, fresh air, rain etc. for their daily life. This village located on the peak of Mount Kulen, if there were no trees or forest, the heat would be higher than the low land. If there was no forest, what would happen for the Siem Reap river? (Men's focus group, Svay Leu District)

4   Discussion

According to the respondents, ACHA project investments in water management infrastructure reduced flood frequency and impact. This logically provided for higher agricultural productivity (e.g. the recorded increase in the value of vegetable production) and contributed to lower levels of severe food insecurity and debt, suggesting significant benefits of improved water management. We did expect benefits to arise, but had not formulated expectations about the scale of benefits. The fact that respondents reported both decreased and increased flood frequency, and decreased and increased impacts, suggests that respondent bias was unlikely to have occurred (i.e. it does not indicate that respondents answered questions according to what they thought interviewers wanted to hear). Data are also corroborated by key informant interviews with APSARA staff (wherein flooding frequency and impact did vary across the Park). We acknowledge that the 3-year time frame for assessment (limited to ACHA project time frame) may not provide a sufficient time period to detect impact.

The benefits of effective SEPL management (i.e. maintenance of water services) clearly support agricultural productivity and improved health outcomes in the lower part of the catchment through food security. While protection of the Park has enabled regulation on deforestation within its boundaries, it has also attracted tourists and most likely resulted in increased population growth in Siem Reap,[9] adding pressure on groundwater resources needed to support the temples. This project demonstrates that while synergies between biodiversity protection and economic development exist that can strengthen SEPL management, new challenges may be created whose mitigation requires a broader landscape focus.

Focus group data from the upper catchment contributed to the understanding of the linkages between water, loss of forests, and human health. Discussions about forest change beyond the Park boundary, compared to the benefit of forests protected within the Park, provide three clear lessons for SEPL management. Firstly, a loss of

---

[9] The rationale for this comment is based on the correlation between population and tourism growth. Between 2008 and 2019, the population of Siem Reap Province grew by 12.3% (Population Census data). Over the same time frame, tourism grew by 149.1% (Ministry of Tourism data on arrivals to Siem Reap by air).

forest coverage impacts human health. Some of these impacts are direct (e.g. shade trees), whereas others are indirect. Indirect changes include a change in land use from forest to rice, cassava, and cashew nut cultivation. The negative impacts of these crops on soil fertility often result in increased rates of agricultural input use (Mahanty & Milne, 2016). Increased health risks are attributed to the use of agricultural inputs. The respondents' awareness of these linkages bodes well for improved SEPL management in the future. Discussions about pathways forward indicated that alternatives do exist that improve water quality and ameliorate health impacts (e.g. traditional composts). Health impacts are particularly important for pregnant women who are more vulnerable, and for women in general who are more susceptible to increases in temperature (WHO, 2014).

Secondly, the loss of biodiversity and land-use changes are clearly associated with development pressure, with participants shifting to higher value cash crops, and complaining about their ability to control the use of forests by "outsiders". Cash crops such as cassava are low labour-intensive, enabling migration of family members for part of the year to supplement household income in the country of origin of migrants (Eliste & Zorya, 2015). Studies from Siem Reap (e.g. Jacobson et al., 2019; Jacobson, 2020) indicate that migration results in a loss of agricultural labour. Losses of agricultural labour can result in less intensive production, or smaller production areas, which can result in lower total production and food availability. This can further exacerbate existing food insecurity for remaining members, in addition to other reported impacts on human health, such as increasing violence against women and children. Outsider use of forests is indicative of the rural-urban poverty divide, whereby the urban poor in Cambodia improved their share of disposable income by nearly twice as much as the urban rich between 2009 and 2017, whereas the rural poor improved it by only 30% as much as the rural rich.[10] Illegal and "outsider" use of forests is an option for rural poor to advance themselves; restricting forest degradation therefore requires carefully targeted development initiatives for this group. Identifying and working with vulnerable groups who engage in deforestation in meaningful ways remain a critical development challenge.

Lastly, the potential impacts of any future deforestation in the upper catchment, coupled with climate change, are expected to result in water shortages and affect temple stability in the Park (Chim et al., 2021). Likewise, any further deforestation in the upper catchment is predicted to also affect the economic benefits from tourism (Chim et al., 2021). This could severely affect tourism benefits associated with the Park, estimated at 4.9 billion USD in 2019 (Ministry of Tourism, 2019). It would also impact the ability to sustain the cultural service benefits associated with the ACHA project in the Park, and could affect the ability to sustain food security benefits within the Park due to the impact on groundwater reserves.

Future challenges to SEPL management include the accelerating pressures of climate change, alongside livelihood development placing further pressure on water

---

[10]Based on the analyses of National Institute of Statistics CSES data on disposable incomes for 2009–2017.

resources. Further lower flows in the Mekong River are anticipated to undergo proposed hydroelectric development (Hecht et al., 2019); this will affect the flood pulse in the Tonle Sap lake, reducing groundwater levels. This compounds changes that result from extractive uses of water. The ability to sustain ACHA project benefits therefore depends on additional interventions that address the concomitant impacts of development, climate change, as well as unknown impacts of the COVID-19 pandemic. Tourism during the pandemic has all but ended, declining by 95.5% between June 2019 and June 2020,[11] with an estimated forgone annual provincial revenue of 1.3 billion USD.[12] The COVID-19 pandemic has also resulted in many migrants returning home, with remittance losses for many; for example, reported rates of migration in Siem Reap province prior to the pandemic were up to 46% of households, with an average remittance of 284 USD (Jacobson et al., 2019); loss of remittances is also likely to affect debt sustainability, and places increased pressures on food availability. These socio-economic challenges are likely to hinder the ability to maintain biodiversity and the benefits derived from ACHA in the future. In recent years, the physical extent of ancient Angkorian water management infrastructure beyond existing protected area boundaries has been studied by scientists (Evans, 2016), although knowledge of it likely persisted through traditional mechanisms. Continued investment in its reinvigoration beyond the Park boundaries is therefore likely to support the whole of landscape management.

Addressing the links between environmental health and human well-being requires recognition that management interventions might need to occur at locations different from where the impacts of poor management are experienced. For example, the impacts of forest degradation are experienced in the Park in terms of lower groundwater levels. However, the most important sites for action are probably in the upper catchment where deforestation is higher, and the mid-catchment where groundwater recharge can be maximised through water-efficient agriculture. In addition to the obvious focus on water, we identified four other strategic directions to enhance environmental health and human well-being outcomes through sustainable management of landscapes.

First, we advocate for community-based natural resource management (CBNRM) to enhance SEPL management in the upper catchment, and complement existing management of the Park in the lower catchment. CBNRM recognises that environmental protection requires the survivability of communities as stewards of the natural environment. Criteria for successful CBNRM include (1) local recognition of problems and initiation of projects, (2) economic incentives for environmental management, (3) alternative livelihoods that combat the opportunity costs of environmental protection, (4) autonomy of decision-making, and (5) capacity development (Measham & Lumbasi, 2013).

---

[11] Based on Cambodian Ministry of Tourism arrival data, reported monthly.

[12] Using figures of the total value of tourism reported by the minister adjusted by share of arrivals. Reference is made to https://www.phnompenhpost.com/business/cata-president-looking-ahead-revival-tourism-industry.

Our analysis demonstrated that community and government staff acknowledged the importance of ecosystem services to livelihoods and well-being (criterion 1). Economic incentives for protection exist as evidenced from economic analysis of water infrastructure management in the Park (criterion 2). However, it is unclear whether these will outweigh the opportunity costs of further actions to limit land-use change and protect forests (criterion 3). Payment for Ecosystem Services (Redford & Adams, 2009) could reduce the rate of land-use change, but the benefit distribution as well as the identification of appropriate buyers of services will be critical. The opportunity cost balance could also be addressed through a more nuanced understanding of vulnerability. This will ensure that even those who are contributing to negative impacts, but doing so inevitably due to their vulnerability, could benefit from development projects so as to be compensated for missed opportunities and ensure that their livelihoods are sustainable. This will be particularly important given the emerging economic impacts of the COVID-19 pandemic. For example, those from outside of forest communities may be more vulnerable than those within, leading them to engage in forest clearing. In development projects, these people, as well as people within forest communities, need to be considered as beneficiaries.

For autonomy (criterion 4), some mechanisms already exist that could be strengthened to protect forests. Management committees for Community Protected Areas and Community Forestry Areas (examples of which exist in the upper catchment) have the ability to set regulations on forest use, but lack resources and a mandate to address regulation breaches. Legal provisions also exist that could support community-based management of water, but these have had limited application (Mak, 2017). Capacity development is also needed, as identified by community members (criterion 5). We propose community-based monitoring of disaster risk reduction and environmental health (e.g. Rainforth & Harmsworth, 2019) as one mechanism for incorporating and improving knowledge and autonomy in management.

Second, we also advocate for nature-based solutions (NBS). NBS specifically promote nature as a means of mitigating climate change and adapting to it (Nesshover et al., 2017). Solutions relevant to the Park include actions that strengthen ecosystem service provision while providing alternative livelihoods. For example, riparian planting could reduce erosion of river banks. It also has the potential to provide food (fruit trees) and jobs in nursery production, as well as shade trees to reduce heat stress. Alternatively, instream structures could be built that slow the speed of water travel across the landscape and increase production yields through soil moisture retention. Health impacts could be ameliorated through the promotion of organic compost instead of chemical inputs. This could take advantage of existing aquatic plants, and potentially improve production returns through organic certification.

Third, we advocate for economic diversification. Economic diversification is considered critical for resilience building (Davidson et al., 2016), and lessening the risk of livelihood failure during times of uncertainty and crisis. Smallholder reliance on one or two key crops heightens risks associated with climate change. Market supply-chain development and careful selection of alternatives in horticultural production will be important, given the market competitiveness of imported fruits and vegetables in Cambodia (World Bank, 2015). Unique heritage crops such as tiger-hand potato, and high-value land-efficient crops such as pepper and moringa, could provide niche market opportunities.

Lastly, we advocate for improved disaster preparedness. Disaster management in Cambodia is in its infancy (ESCAP, 2016). Understanding the temporal-spatial dimensions of disasters within the catchment, and facilitating preparedness for sudden and slow on-set events (i.e. emergency responsiveness, preparedness, and adaptation), will enhance resilience. This includes planning for emergency water and food reserves (i.e. rice), adaptation in crop varieties and crop choices, and preparedness for emergency livelihoods (e.g. tuber crop stems and rice seed).

5 Conclusion

Our case study highlights the potential benefits of effective SEPL management from re-invigorating ancient water management systems—benefits valued at over nine million USD over four years in Angkor Park. Protection of forests in the Park and water management arguably supported improved livelihoods. Improved hydrological and biophysical monitoring and regular socio-economic and health data collection would enable more detailed quantitative assessment of the relationship between changes in biodiversity, health, and sustainable development.

Sustaining these benefits into the future is a challenge. Increased pressure from proximate sources (e.g. tourism in Siem Reap), global changes (e.g. climate change), and rural-urban poverty divide that is contributing to land-use change are examples of ongoing pressures on water resource management. Population and developmental pressures coupled with climate uncertainty and changes in water availability mean that a water deficit is likely to remain, or even worsen, in the future. This provides a trade-off between tourism and water management. Tourism reinforces the need for conservation and protection of forests. However, it also adds extractive pressure on water resources required to maintain the temples. As a result, consideration of the socio-economic production landscape at the catchment scale is needed to facilitate adaptation and long-term sustainability—the same scale as ecological processes that influence hydrological dynamics.

Our research demonstrates that effective SEPL management of the entire Siem Reap catchment is needed, even if the socio-economic dimensions are complex. While community members are aware of environmental deterioration and its potential impacts on the Angkor Archaeological Park, their ability to address forest and land degradation cannot be decoupled from the need for poverty reduction. We highlight potential solutions that recognise the dependence of communities on natural resources, and generate benefits for communities and nature. Upstream investment in water infrastructure, community-based management of forest and water resources, and nature-based solutions such as riparian food forest planting are some examples. Investments in disaster preparedness and targeting of activities to the most vulnerable will be critical to removing disincentives to adaptation that result from absolute poverty.

Acknowledgements The research presented was supported with funding from the Ministry of Foreign Affairs and Trade, Manatū Aorere, New Zealand, through Angkor Community Heritage and Advancement project, and the Angkor Landscape Management Strategy. Part of this work included economic valuation of water management conducted by Jacobson, Professor Ken Hughey, and Emeritus Professor Ross Cullen (Lincoln University, New Zealand), with secondary data provided through partnership with the University of Battambang (Cambodia), New Colombo Plan (DFAT, Australia), and Food and Agriculture Organization of the United Nation's Life and Nature project. Our sincerest thanks to community members and officials for their participation, and our teams for support with data collection.

References

ADB. (2012). *Flood damage emergency reconstruction project, preliminary damage and loss assessment.* Asian Development Bank.

Carletto, C., Zerra, A., & Banjerjee, R. (2013). Towards better measurement of household food security: harmonizing indicators and the role of household surveys. *Global Food Security, 2*(1), 30–40.

Chen, F., Guo, H., Ma, P., Lin, H., Wang, C., Ishwaran, N., & Hang, P. (2017). Radar interferometry offers new insights into threats to the Angkor site. *Science Advances, 3*(3), e1601284.

Chim, K., Tunncliffe, J., Shamseldin, A., & Ota, T. (2019). Land use change detection and prediction in upper Siem Reap river, Cambodia. *Hydrology, 6*(3), 64.

Chim, K., Tunnicliffe, J., Shamseldin, A., & Sarun, S. (2021). Sustainable water management in the Angkor temple complex, Cambodia. *SN Applied Sciences*, *3*(74), viewed 20 September 2021. Retrieved from https://link.springer.com/content/pdf/10.1007/s42452-020-04030-0.pdf.

Davidson, J. L., Jacobson, C., Lyth, A., Dedekorkut-Howes, A., Baldwin, C. L., Ellison, J. C., Holbrook, N. J., Howes, M. J., Serrao Neumann, S., Singh-Peterson, L., & Smith, T. F. (2016). Interrogating resilience: toward a typology to improve its operationalization. *Ecology and Society*, *21*(2), art. 27, viewed 20 September 2021. Retrieved from https://doi.org/10.5751/ES-08450-210227.

ESCAP. (2016). Country report of Cambodia disaster risk reduction. In *Regional capacity development workshop: mainstreaming DRR in sustainable development planning*, New Delhi, viewed 28 January 2021. Retrieved from https://www.unescap.org/sites/default/files/Cambo dia%20Country%20Presentation.pdf.

Eliste, P., & Zorya, S. (2015). *Cambodian agriculture in transition: opportunities and risks.* World Bank Group.

Evans, D. (2016). Airborne laser scanning as a method for exploring long-term socio-ecological dynamics in Cambodia. *Journal of Archaeological Science, 74*, 164–175.

Frappart, F., Biancamaria, S., Normandin, C., Blarel, F., Bourrel, K., Aumont, M., Azemar, P., Vu, P.-L., Toan, T. L., Lubac, B., & Darrozes, J. (2018). Influence of recent climatic events on the surface water storage of the Tonle Sap lake. *Science of the Total Environment, 636*(15), 1520–1533.

Google. (2021). Google Maps [Angkor Archaeological Park and Siem Reap River Catchment], viewed 20 June 2021. Retrieved from https://www.google.com/maps/d/edit?mid=1 iCUTESUypcV-3z1KHY4FOvjx6onUsbw2&usp=sharing.

Hang, P., Ishwaran, N., Hong, T., & Delanghe, P. (2016). From conservation to sustainable development—a case study of Angkor World Heritage site, Cambodia. *Indian Journal of Environmental Heath, 5*(3), 141–155.

Hecht, J. S., Lacombe, G., Arias, M. E., Dang, T. D., & Piman, T. (2019). Hydropower dams of the Mekong river basin: a review of their hydrological impacts. *Journal of Hydrology, 568*, 285–300.

Jacobson, C., & Smith, J. (2020). *Landscape Voices: a rapid assessment of the perceptions into Angkor Landscape Management*. Live & Learn.

Jacobson, C., Crevello, S., Chea, C., & Jarihani, B. (2019). When is migration a maladaptive response to climate change? *Regional Environmental Change, 19*(1), 101–112.

Jacobson, C., Cullen, R., & Hughey, K. F. D. (2018). *Economic valuation of water management in association with Angkor Community Heritage Advancement project*. Report to Apsara Authority and Live and Learn.

Jacobson, C., & Seng, R. (2018). Unpublished data on food security and agricultural change in Lovea Kriang Commune, Siem Reap. University of Battambang, Cambodia, and University of the Sunshine Coast.

Jacobson, C. (2020). Community climate resilience in Cambodia. *Environmental Research, 186*, 109152.

JICA. (2004). *JICA Guideline for project evaluation*, viewed 20 September 2021. Retrieved from https://www.jica.go.jp/english/our_work/evaluation/tech_and_grant/guides/guideline.html.

Kirsch, H. (2010). Watershed inventory Siem Reap, Cambodia: A combination of social and natural science methods. *Pacific Geographies, 34*, 9–14.

Nakau. (2020). Options assessment: Payment for ecosystem services in the Angkor Landscape Management Strategy. Unpublished report to Live & Learn Environmental Education.

Mahanty, S., & Milne, S. (2016). Anatomy of a boom: Cassava as a 'gateway' crop in Cambodia's North-eastern borderland. *Asia Pacific Viewpoint, 57*(2), 180–193.

Mak, S. (2017). Water governance in Cambodia: from centralized water governance to farmer water user community. *Resources, 6*(2), 44.

Measham, T. G., & Lumbasi, J. A. (2013). Success factors for community-based natural resource management (CBNRM): lessons from Kenya and Australia. *Environmental Management, 52*, 649–659.

Ministry of Tourism. (2019). *Tourism year book*. Ministry of Tourism.

Nesshover, C., Assmuth, T., Irvine, K. N., Rusch, G. M., Waylen, K. A., Delbaere, B., Haase, D., Jones-Walters, L., Keune, H., Kovacs, E., Krauze, K., Kulvik, M., Rey, F., van Dijk, J., Vistad, O. I., Wilkinson, M. E., & Wittmer, H. (2017). The science, policy and practice of nature-based solutions: An interdisciplinary perspective. *Science of the Total Environment, 579*, 1215–1227.

National Institute of Statistics. (2017). *Cambodia Socio-Economic Survey 2017*. Ministry of Planning.

National Institute of Statistics. (2009). *Cambodia Socio-Economic Survey 2009*. Ministry of Planning.

Rainforth, H., & Harmsworth, G. (2019). *Kaupapa Māori freshwater assessments: A summary of iwi and hapū-based tools, frameworks and methods for assessing freshwater environments*. Perception Planning Ltd..

Redford, K. H., & Adams, W. M. (2009). Payment for ecosystem services and the challenge of saving nature. *Conservation Biology, 23*(4), 785–787.

Seenprachawong, U. (2006). *Economic Valuation of Cultural Heritage: A Case Study of Historic Temples in Thailand*. Economy and Environment Program for Southeast Asia, viewed 21 January 2021. Retrieved from http://econ-test.nida.ac.th/en/index.php?option=com_content&view=article&id=532%3Aeconomic-valuation-of-cultural-heritage-a-case-study-of-historic-temples-in-thailand&catid=29%3Apublication-33&Itemid=117&lang=en.

Tuan, T. H., Seeprachawong, U., & Navrud, S. (2009). Comparing cultural heritage values in South East Asia: possibilities and difficulties in cross-country transfers of economic values. *Journal of Cultural Heritage, 38*, 1–13.

Wales, N. (2020). An examination of forest cover change at Angkor, Cambodia, using satellite imagers, interview and interpretation of historical events. *Applied Geography, 122*, 102276.

WHO. (2014). *Gender, Climate Change and Health.* World Health Organization.

World Bank. (2015). *Cambodian agriculture in transition: opportunities and risks.* Economic and Sector work report no. 96308-KH, World Bank, Washington DC, USA.

The opinions expressed in this chapter are those of the author(s) and do not necessarily reflect the views of UNU-IAS, its Board of Directors, or the countries they represent.

Open Access This chapter is licenced under the terms of the Creative Commons Attribution 3.0 IGO Licence (http://creativecommons.org/licenses/by/3.0/igo/), which permits use, sharing, adaptation, distribution and reproduction in any medium or format, as long as you give appropriate credit to UNU-IAS, provide a link to the Creative Commons licence and indicate if changes were made.

The use of the UNU-IAS name and logo, shall be subject to a separate written licence agreement between UNU-IAS and the user and is not authorised as part of this CC BY 3.0 IGO licence. Note that the link provided above includes additional terms and conditions of the licence.

The images or other third party material in this chapter are included in the chapter's Creative Commons licence, unless indicated otherwise in a credit line to the material. If material is not included in the chapter's Creative Commons licence and your intended use is not permitted by statutory regulation or exceeds the permitted use, you will need to obtain permission directly from the copyright holder.

Chapter 8
Are the Skiing Industry, Globalisation, and Urbanisation of Alpine Landscapes Threatening Human Health and Ecosystem Diversity?

Andrea Fischer, Henriette Adolf, and Oliver Bender

Abstract The Jamtal Environmental Education Centre is a joint effort of the local communities of Galtür and Ischgl and the Alpinarium museum, dedicated to high mountain livelihoods and landscapes. For this study, we compiled available scientific evidence and personal views in the two communities on the co-evolution of human health and the biodiversity of local ecosystems. Main sources are historical records and maps, chronosequencing in the glacier forefields, and an analysis of contemporary land cover and glacier changes. In both communities, a large part of the area has remained unused since the start of the records in 1857. While the glacier area has shrunk by 70% since then, the forest area has increased as a result of changing land use and climate. Chronosequencing reveals that the glacier forefields are refugia for cold-adapted species under pressure from climate warming. Although land cover has changed, no type of land use recorded in the historical data has disappeared completely. While health services and infrastructure are thought to be sufficient, interviewees saw the largest potential for improvement in today's lifestyle. Traditional practices involving usage of herbs or food culture, for example related to *Gentiana punctata*, are still alive and important for the communities.

Keywords European Alps · Tourism · Biodiversity · Health · Museum · Silvretta

1 Introduction

Both conservation and sustainable use of mountain biodiversity can ensure the health of humans and the ecosystem and increase the resilience of socio-ecological systems. In this sense, mountain municipalities are promoting their landscapes as

A. Fischer (✉) · H. Adolf · O. Bender
Institute for Interdisciplinary Mountain Research of the Austrian Academy of Sciences, Innsbruck, Tyrol, Austria
e-mail: andrea.fischer@oeaw.ac.at

© The Author(s) 2022

M. Nishi et al. (eds.), *Biodiversity-Health-Sustainability Nexus in Socio-Ecological Production Landscapes and Seascapes (SEPLS)*, Satoyama Initiative Thematic Review, https://doi.org/10.1007/978-981-16-9893-4_8

destinations for sports, wellness, and even health tourism. The upper Paznaun Valley in the Austrian Alps is one of the highest settlements in the Eastern Alps. With glaciers and badlands like steep rocky mountain faces occupying 69% of the area of the community of Galtür and 37% of the community of Ischgl, only 1% and 5% of the area, respectively, is inhabitable. Therefore, the majority of the land is not used directly, despite the evidence of human impact in pollen profiles for the last millennia.

The landscape we find there today is shaped by various geomorphological processes, ongoing agricultural use, and modern infrastructure such as traffic, energy production, and tourism. The community of Galtür and the Alpinarium Society founded a museum, the Alpinarium (https://www.alpinarium.at), which came into operation in 2003. This *Erlebnismuseum* (experiential museum) is dedicated to livelihoods in high mountain regions. In 2013, the Jamtal Environmental Education Centre was established as a joint venture between the Alpinarium, the community of Galtür, and the Institute for Interdisciplinary Mountain Research of the Austrian Academy of Sciences to improve the accessibility of scientific results and initiate a discourse with the public. The main target audience has been pupils, who can spend a week at a mountain hut and experience high mountain nature and climate as a first step to come to a deeper understanding of mountain landscapes based on their own experience. Now the Environmental Education Centre has gained more cooperation partners, like the local guides who help to extrapolate the Alpinarium museum from the building into the backyards and outdoors. But the major impulse towards transdisciplinarity arose from the extensive exchange between the cooperation partners, so that the needs expressed by the communities also shape the upcoming scientific studies.

Of the data compiled for the Environmental Education Centre, this chapter pulls together those scientific data related to biodiversity and health. It shows that a look back in time can help to understand current environmental changes and their consequences for local people. Located near the border between Austria, Switzerland, and Italy, a nodal point of cross-Alpine mobility, the upper Paznaun Valley has been inhabited and farmed for millennia. During industrialisation, with decreasing trade, the prosperity of the region fell, forcing families to send their children abroad for work. Tourism started in the late nineteenth century, changing (but not entirely replacing) agricultural land use and increasing the mobility and exchange of goods, people, and ideas. Consequently, poverty was alleviated, the provision of affordable (clean) energy improved, and transport infrastructure expanded throughout this high-alpine environment, with large investments in avalanche mitigation measures, all of it deeply altering the socio-ecological production landscape. Changes in agricultural land use led to decreasing biodiversity. From the mid-nineteenth century onwards, glaciers retreated, becoming one-third of their former size. The rapid succession of plants in the ice-free former glacier areas has created new ecological niches.

This study is focused on the co-evolution of agricultural and touristic land use, and related changes in social settings. We tackled the former through a detailed analysis of geodata and investigated the latter by interviews with stakeholders of the two communities of Ischgl and Galtür. The landscape is permanently changing as a result of both natural processes and (changing) cultural practices. We analyse the use of agricultural areas in particular, as this has a high impact on biodiversity in the settlement area.

Within the larger framework of the UN 2030 Sustainable Development Goals (SDGs), this study illustrates that the communication of pathways of sustainable development in the past and present is a necessary endeavour to ensure the balanced development of healthy and biodiverse socio-ecological production landscapes.

1.1   Geographical Information

The communities of Ischgl (1377 m asl) and Galtür (1600 m asl) are the highest villages in the Paznaun Valley, right at the national border with Switzerland (Fig. 8.1 and Table 8.1), in the Austrian state of Tyrol. The highest point of the two communities is the Fluchthorn peak (3399 m asl).

Located right at the margin of permanently inhabitable areas, the villages have a rough climate with an annual mean temperature of 2.7 °C (Galtür 1957–2000) and high annual precipitation (1013 mm; Fischer et al. 2019). In recent years, monthly temperature means have exceeded past averages by up to about 4 °C. Average height of winter snow cover is among the highest in Austria, and the topography-induced precipitation in the case of north-westerly flows leads to high precipitation rates. Now the glaciers are receding extremely rapidly (Fischer et al. 2021), with a potential total deglaciation in a few decades. This could have major effects on inhabited areas due to glacier outburst floods during the rapid retreat and sediment erosion and rock falls after the loss of ice (Kääb et al. 2005). The high number of rock glaciers in the area (Krainer & Ribis 2012) holds additional potential for destabilisation during the current climate change.

The area has been inhabited for at least 6000 years (Dietre et al. 2014), with a history of migration and adaption of different ethnolinguistic groups from German-speaking Upper Valais, Rhaeto-Romance Switzerland, Tyrol, and Vorarlberg (cf. Bender & Haller 2017). The specific groups brought with them greatly differing agricultural practices, making the region a rich repository of diverse cultural practices (Gemeinde Galtür 1999).

The glaciers are located close to pastures, and the trails used for trading cattle and goods were reportedly unpassable due to the glacier advance during the Little Ice Age (Huhn 1997). This caused major problems, as until the sixteenth century, Galtür and Ischgl were part of the community of Ardez in the Engadine Valley in the Grisons (now part of Switzerland), and the area had been used as summer pasture as was common along the main Alpine ridge in the Eastern Alps.

Working migration has involved the outgoing seasonal working migration of poor alpine peasants' children, mostly 6–14 years old, to big farms in the German (Suebian) foreland (so-called *Schwabenkinder*) continuing from the fifteenth to the early twentieth century (Ulmer 1943), and currently includes the presence of incoming seasonal workers mainly from Eastern Europe (Bender 2015).

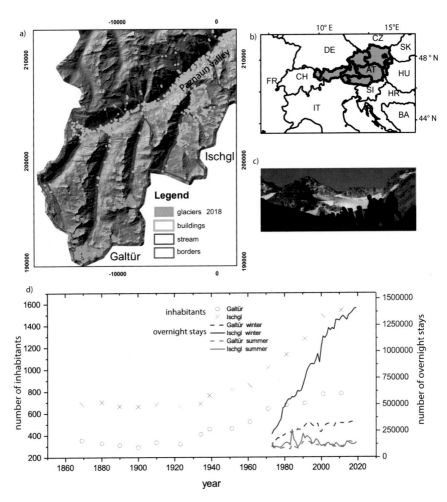

Fig. 8.1 Site map (**a**) and location of the communities (in red) in Austria (AT) (**b**) of the communities of Ischgl and Galtür in upper Paznaun, with (**c**) pupils in action with the Environmental Education Centre in Jamtal on Jamtalferner (photo by Andrea Fischer), and (**d**) total population and overnight stays (data source: Federal Administration of Tyrol)

From about 1870, the first touristic infrastructure began to be constructed, mainly mountain huts. Slowly, summer tourism also developed down in the valley, but it was not until the 1960s that cable cars were built—60 years after ski tourism reached the valley. In the 1970s, earnings from winter tourism exceeded those from summer tourism, and have been increasing since then, more strongly in Ischgl than in Galtür. In 2019 both municipalities together had ca. 441,000 tourist arrivals and over two million overnight stays (Land Tirol 2021).

In 2020, a population of 766 lived in Galtür on the 6 km^2 of inhabitable area of a total municipal area of 121 km^2. The community of Ischgl had 1604 inhabitants

Table 8.1 Basic information of the study area

Country	Austria
Province	Tirol
District	Landeck
Municipality	Ischgl, Galtür
Size of geographical area (hectare)	22,400
Number of direct beneficiaries (persons)	776 + 1604
Number of indirect beneficiaries (persons)	776 + 1604
Dominant ethnicity(ies), if appropriate	n.a.
Size of the case study/project area (hectare)	22,400
Geographic coordinates (latitude, longitude)	46°59′17.76″N, 10°14′55.65″E

(5.2 km^2 inhabitable area, 103 km^2 total area). The population in Ischgl and Galtür slightly decreased from 355 inhabitants in 1869 for five decades, with a strong reversal of the trend after the census of 1923; however, the resulting average increase for 1870–1940 amounted to only 0.25% a year, while after the Second World War population increased significantly by 1.25% a year (Statistik Austria 2021, Fig.8.1d).

This population increase was due to a positive population change (birth surplus, e.g. 2001–2011: +144), which over both time periods, before and after the Second World War, exceeded the negative spatial population change (migration balance, e.g. 2001–2011: −87). During the last 50 years, demographic change has led to a significant ageing of the population. While the age group of 0–14 years halved, the 15–29 group stagnated, and all groups for 30+ years increased by around one-third. This trend is a result of a strong decrease in birth rate and the increasing outmigration of young people, mostly in search of faraway higher education opportunities and related workplaces. On the contrary, for the period 2001–2011, about 300 persons of the 50–74 age group in-migrated from more distant places (other political districts or abroad) (all data from GALPIS), making the upper Paznaun Valley a typical destination for amenity and retirement residences (Bender & Kanitscheider 2012; Bender 2015).

Livelihoods in the valley have also exhibited distinct change (Table 8.2, Figs. 8.1 and 8.2). While the number of workplaces in the agrarian sector became nearly insignificant, employment in the traffic and tourism sectors increased substantially, especially in Ischgl. Likewise, from 1971 to 2011, the proportion of commuters doubled (Fig. 8.2). In 2011, 24% of employed persons living in Ischgl and Galtür were commuting out to other municipalities, and 38% of the persons employed in the communities were commuting in. Especially for Ischgl, the number of persons commuting in was three times that of those commuting out (all data from GALPIS). Especially due to commuters, multilocal living (357 second residences in 2011 for both municipalities), and tourists, mobility has become very high, up to a degree which is typical for winter sport resorts in Tyrol in the uppermost locations of the tributary valleys (Bender & Borsdorf 2014).

Table 8.2 Change of workplaces by economic sectors in the municipalities of Galtür and Ischgl 1971–2011 (source: GALPIS with temporally harmonised data after the Austrian classification of economic activities ÖNACE)

	Galtür (1971)	Galtür (2011)	Ischgl (1971)	Ischgl (2011)
Workplaces (total number)	235	235	360	864
Agriculture (%)	23.4	5.1	30.6	1.2
Production of goods (%)	4.7	5.1	6.7	2.1
Supply of water and energy (%)	6.4	0.0	1.7	0.8
Construction (%)	5.1	11.5	5.3	0.2
Commerce and repair (%)	5.5	13.2	5.6	11.5
Traffic and communication (%)	6.8	10.6	13.9	28.0
Tourism (%)	28.9	31.9	18.3	37.4
Banking and insurance; housing and business-related services (%)	0.9	8.5	0.8	10.8
Other public and private services (%)	17.9	14.0	14.4	8.1
Unknown/not classified (%)	0.4	0.0	2.8	0.0

1.2   Human-Nature Interaction

Human-nature interaction in the communities over the long term is recorded in pedological and palynological records (Dietre et al. 2014) and shows the strong impact of human presence on biodiversity, e.g. by grazing cattle or managing forests, and wildfires over at least the last six millennia. Cultivation of land and land use reached up to the glaciers even as late as the 1870s, with the majority of Neolithic to Bronze Age activities taking place above today's tree line (Kutschera et al. 2014). The federal maps of the Habsburg Monarchy allow the first detailed insight into the spatial distribution of land cover at one point of time. The first land register, the *Franziszeischer Kataster* (e.g. Scharr 2018), was drawn up to assign the amount of taxes to be paid to reflect different types of land use/cover (for classes, see Fig. 8.2). In 1963 Böhm (1970) mapped land cover in the area again in a detailed study. Although glaciers are much smaller now, and climatic conditions much better, the area used for grazing cattle is much smaller today—partly the effect of switching from sheep and goats to cows. Today, irrigation of dry meadows is not as important as it was before 1950, and the forest is reconquering the area up to the actual glacier termini (Fischer et al. 2019). The now ice-free areas formerly covered by glacier tongues are discussed as refugia for species under pressure by thermophilisation, i.e. species adapted to warm conditions grow in areas where species adapted to cold conditions used to grow (Pauli et al. 2012).

Early tourism started up in the mountains, with the building of huts by Alpine Clubs (von Pfister 1911), not down in the valley. There, tourists were not very appreciated, as, in contrast to travellers and traders, they were thought of as being demanding and reluctant to pay for services. This view has definitely changed. In contrast to other regions, like neighbouring Silbertal in Montafon, where spas have

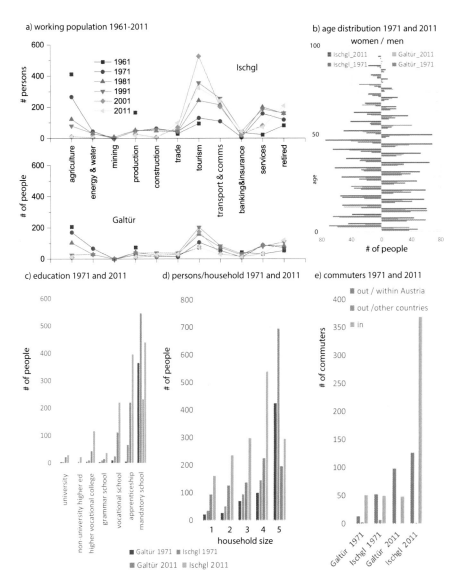

Fig. 8.2 Changes in working population by sector (**a**), age distribution (**b**), education (**c**), number of persons per household (**d**), and commuters (**e**) of the communities of Ischgl and Galtür in upper Paznaun (prepared by author, data source: Federal Administration of Tyrol)

been reported as early as 1616, neither early health tourism, such as visiting healing baths, nor pilgrimages took place in upper Paznaun.

Only 0.18% of the area of Galtür and 2.8% of the area of Ischgl are used for ski runs and cable cars (Table 8.3), but most of this area is also part of pastures. Over the years, the construction of cable cars has been discussed by the local community, as

Table 8.3 Proportion of area which could be used for ski resorts, area actually used for ski resorts and other sports, and inhabitable area in total numbers and percentages (data source: Federal Administration of Tyrol)

Municipality	Area reserved for skiing in spatial planning		Area used for ski runs		Area for other sports and recreation		Inhabitable area	
	km^2	%	km^2	%	km^2	%	km^2	%
Galtür	2.520	2.079	0.189	0.156	0.006	0.005	1.590	1.312
Ischgl	22.956	22.215	2.823	2.732	0.022	0.022	5.200	5.032

well as on a broader scale. Some projects, like the construction of the Piz Val Gronda cable car in Ischgl in 2013, were carried out despite protests by NGOs. Others were debated and rejected by the locals, like the proposal of a glacier ski resort on Jamtalferner glacier in 1976. Construction of skiing infrastructure is not only based on the decision-making of the locals. Federal and state laws stipulate a complex procedure of approval in terms of water resources, environmental protection, and various other aspects before cable cars can be built. Even so, the effects on nature are debated by a much wider and diverse community (Fig. 8.3).

1.3   Health-Related Issues of Livelihood

Out of many potential scientific approaches towards measuring sustainability, well-being, and health, indicators such as the Years of Good Life (YoGL; Lutz et al. 2021) or the SDG Indicators (United Nations Statistical Commission 2017) for Goal 3, "Ensure healthy lives and promote well-being for all at all ages", try to capture the status quo and make improvements visible. A list of aims and the respective indicators in the framework of the SDGs can be found in Appendix 1. Austria, where the study area is located, is among the most developed countries. This results in high scoring on SDG health indicators related to good health infrastructure such as medical care and hospitals, as well as some low scoring on other health indicators, for example, for alcohol consumption (Fig. 8.4).

Changes in livelihoods in the study area since 1850 are diverse, with industrialisation and recently digitalisation affecting society. Böhm (1970) described various physical, cultural, ecological, and societal aspects of the site at the beginning of the period of rapid economic development ("*Deutsches Wirtschaftswunder*") after the Second World War, which drastically changed livelihoods and the landscape even in the remotest parts of central Europe. Mountain farming was influenced by mechanisation of work, with increasing amounts of machinery and decreasing numbers of workers (Lichtenberger 1965). Figure 8.5 illustrates that the total working hours for a typical farmer were distributed more equally around the 2010s than in the 1960s, as holidays from the secondary occupation (in most cases in tourism) were used for farming work. Figures for the 1960s and the 2010s are based on a survey made by Böhm (1970) in the 1960s and the authors in 2021. In the

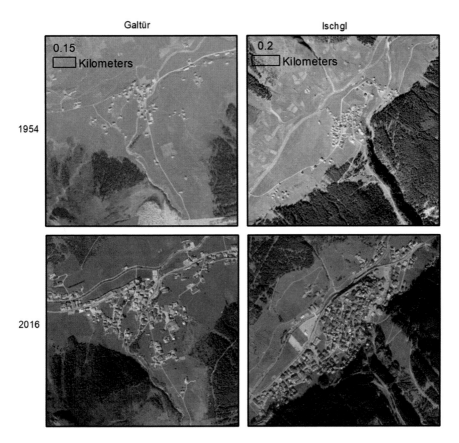

Fig. 8.3 Orthophotos of the communities of Galtür (left) and Ischgl (right) in 1954 and 2016 (data source: Federal Administration of Tyrol). The number of buildings and traffic areas have increased significantly. North of the communities, some of the former agricultural land is occupied by forest, and south of Ischgl, part of the forest has become a ski slope

2010s, 3 h per day were needed for farm work, with 2 weeks of holidays made possible by supportive manpower. The holidays at the end of the skiing season were used for preparatory work, and few days were spent for hay work in July and the second week of September. The cattle are grazed on the mountainside from June 15 to September 10 or 15, with 3 weeks of grazing outside the stable afterwards. Tree felling and other additional work have been excluded from this study. Böhm (1970) listed small bed and breakfast activities as typical businesses complementary to farming in the 1960s. Although this type of business was often run by female family members, empowering them by allowing them to earn their own money (Schmitt 2010), the summer season was part of a very busy farming time. In contrast, in more recent times the period where two occupations overlap is much shorter.

During intensive working periods, e.g. making hay, external workers or family members are needed as extra labour. While in the 1960s farming was the main part of

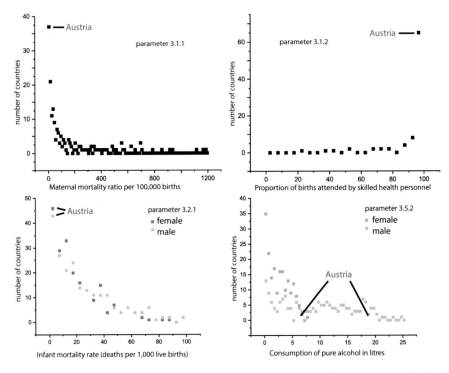

Fig. 8.4 The parameters of the study area, as part of Austria, with all other countries of the world for indicators 3.1.1, 3.1.2, 3.2.1, and 3.5.2 show that the supply of healthcare in Austria is above average, but lifestyle-related health issues such as alcohol consumption clearly have the potential for improvement (data source: United Nations Statistical Commission 2017)

working life, in the 2010s the "secondary" occupations (blue line) are the main and most continuous part of the working life of the farmer, due to the workload being reduced by the mechanisation of farming (data sources: Böhm 1970, author's survey).

2 Description of Activities and Methods

2.1 The Environmental Education Centre

The Jamtal Environmental Education Centre at Jamtalhütte hut was founded in 2013 to foster the transfer of scientific knowledge, mainly on the effects on climate change, to the interested public directly on-site. The community of Galtür and the local Alpinarium museum supported this project. Based on more than 30 years of research on climate and glaciers in the area, the project has come to include more and more disciplines and opened a dialogue between local people and scientists. This

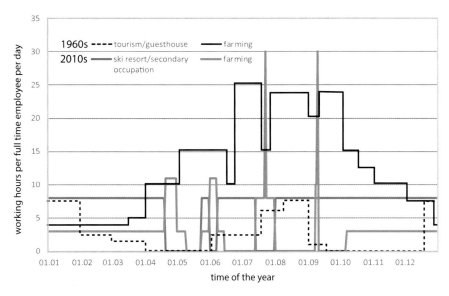

Fig. 8.5 Working time for a typical farmer in the 1960s (solid black line) with additional earnings from small guesthouses (dashed black line) and farming and paid work time in the 2010s (source: Böhm 1970, author's survey)

dialogue comes together in the Alpinarium, where the local community presents high-alpine livelihoods and environment, their changes over time, and their relation to global processes.

Local activities include usage of a flowering plant called a gentian (*Gentiana punctata*), designated as an intangible cultural heritage (UNESCO 2013). The roots of *Gentiana punctata* are collected in a strictly regulated way to produce schnapps used as medicine. Despite being forbidden for a short period after 1991, the practice of collecting roots has been allowed again. Likewise, the traditional and sustainable practice of collecting grassland plants and lichens as described by Zwitter and Rasran (2022) has long been part of the local identity and economy, with the numbers of gentian rising due to the intensification of land use during the Neolithic and later periods. Zwitter and Rasran (2022) show that the usage of medical plants led to an increase in the number of this species. Therefore, a modern restriction on plant usage has no guarantee of preserving the niche of a species in an ecosystem, which is the result of thousands of years of co-evolution of humans and the biosphere. The now joint efforts of locals and scientists aim to tackle past pathways in the co-evolution of nature and society to build a sustainable future. "Living in harmony with nature" implies dynamic action and response from both sides, humans and nature, to cope with pulse and pressure dynamics as societal response to environmental pressure and vice versa (e.g. Collins et al. 2007).

2.2 Analysing Land-Use Change from Historical Maps and Land Cover Types

Repeated data on biodiversity in the area are rare and restricted to single plots. For example, the chronosequencing of the periglacial area of Jamtalferner shows the rapid succession of vascular plants after glacier retreats (Fischer et al. 2019). One of the major drivers of changes in biodiversity is land use, as shown in numerous studies (e.g. Mayer et al. 2009, Brugger & Erschbamer 2012). To get an impression of the spatial distribution of land cover/land use and its changes, we analysed historical maps and modern data of the Federal Administration of Tyrol. The changing extent of glaciers (Fischer et al. 2021) and the area of rock glaciers complement the total picture of land cover/land use and badlands in the study area.

In addition, information on historical and present infrastructure related to human health was compiled, including local health infrastructure.

2.3 Interviews

To reach this study's aims, we carried out interviews with mayors and managers in the local authorities (*Gemeindeamtsleiter*) in February 2021. The interviews were structured and followed a set of open-ended questions (see Appendices A2 and A3).

3 Results

3.1 Land-Use Change

In both communities, the portion of agrarian land (agriculture, pastures) has decreased since 1857 (Figs. 8.6 and 8.7, Table 8.4). In 1857, alpine pastures reached up to the actual glacier areas, but these are not used for grazing cattle today. The largest part of the areas of both communities is badlands, forests, and alpine meadows. In Galtür, large-scale timber production for the salt mine located in Hall reduced the total amount of forest drastically in the nineteenth century. Today the proportion of forest in Galtür is small but on the increase again as a result of abandoning of alpine pastures and of the warming climate raising the treeline. In 1857 traffic routes (footpaths) led to the Swiss border. Today no road across the border exists. Recreational areas are not evident from the 1857 maps. In relative figures, the fraction of both recreational and inhabitable areas in both communities is small. Therefore, most areas in both communities can be considered to be unaffected by direct local human influence. Larger scale pressures like climate change,

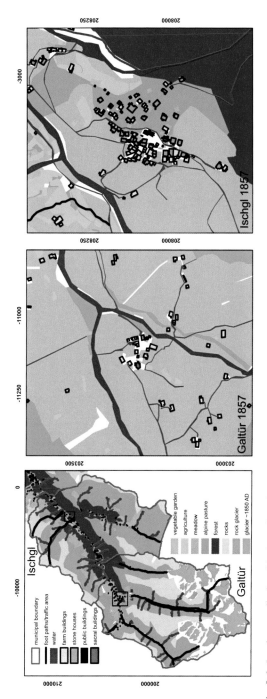

Fig. 8.6 Land use as displayed in the cadastral map *Urmappe/Franziszeische Mappe* of 1857 (map created by the authors based on the cadastral map Urmappe/Franziszeische Mappe)

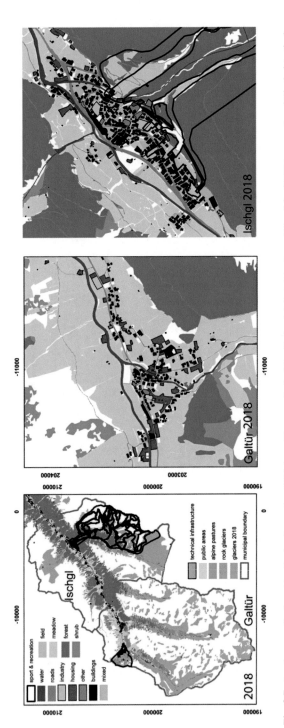

Fig. 8.7 Land use as displayed by the land-use plan (*Flächenwidmungsplan*) of the Federal Administration of Tyrol for 2018 (source: Land Tirol, data.tirol.gv. at 2021)

Table 8.4 Land use in the communities of Galtür and Ischgl in absolute figures. The year 1857 was extracted from the map of the land register of Francis I (*Urmappe* of the *Franziszeischer Kataster*), and other data were provided by the Federal Administration of Tyrol. Although the data sources have been compiled in very different legal frameworks and settings, they allow for interpretation of the general picture of changes, which can also be gleaned from changes in the classifications themselves. For example, the footpaths of the *Urmappe* have only limited economic or legal impact today, as most of the traffic is on roads. Traffic areas were not compiled for every year in the table

	Year	Buildings km²	Agriculture km²	Vegetable gardens km²	Alpine pasture km²	Forest km²	Water km²	Traffic area km²	Other km²
Galtür	1857	0.024	6.009	0.002	46.087	4.657	0.482	0.044	63.930
	1944	0.039	2.935	0.000	45.885	3.917	0.000		68.389
	1973	0.050	3.091	0.000	45.471	3.939	0.448		68.166
	1983	0.056	3.076	0.000	45.471	3.939	0.456		68.167
	1987	0.062	3.070	0.000	45.469	3.939	0.456		68.169
	1991	0.070	3.057	0.000	45.469	3.939	0.457	0.307	68.174
	1995	0.084	2.919	0.001	45.449	4.084	0.460	0.312	67.866
	1998	0.085	2.845	0.025	45.543	4.045	0.457	0.316	68.174
	2001	0.104	2.636	0.100	45.515	4.131	0.460	0.356	68.227
	2002	0.110	2.667	0.097	45.482	4.131	0.460	0.356	68.227
	2005	0.077	2.660	0.103	45.482	4.131	0.462	0.360	68.264
	2014	0.106	2.532	0.100	33.102	14.829	1.603	0.478	69.381
	2015	0.106	2.531	0.100	33.102	14.829	1.603	0.478	69.381
	2020	0.109	2.527	0.109	33.047	14.833	1.608		68.995
Ischgl	1857	0.050	13.437	0.014	51.772	21.620	0.333	0.084	16.205
	1944	0.068	3.847	0.000	61.851	19.659	0.000		18.002
	1973	0.075	4.292	0.000	61.829	19.285	0.318		17.628
	1983	0.120	4.240	0.002	61.826	19.248	0.325		17.629
	1987	0.139	4.184	0.002	61.819	19.238	0.326		17.681
	1991	0.149	4.170	0.002	61.819	19.238	0.328	0.362	17.682
	1995	0.130	4.199	0.016	61.803	19.224	0.328	0.368	17.260

(continued)

Table 8.4 (continued)

Year	Buildings km²	Agriculture km²	Vegetable gardens km²	Alpine pasture km²	Forest km²	Water km²	Traffic area km²	Other km²
1998	0.140	4.122	0.056	61.803	19.213	0.336	0.405	17.661
2001	0.250	3.504	0.216	61.793	19.497	0.338	0.442	17.731
2002	0.255	3.490	0.211	61.793	19.455	0.338	0.443	17.786
2005	0.132	3.482	0.221	61.790	19.449	0.342	0.452	17.924
2014	0.231	3.789	0.137	37.246	23.816	1.012	0.750	36.087
2015	0.231	3.790	0.138	37.246	23.816	1.012	0.750	36.086
2020	0.235	3.780	0.142	37.404	23.768	0.995		37.014

pesticides, and accumulation of microplastics by rain and snow, however, occur. Areas set free by the shrinking Alpine glaciers provide new refugia and niches for vascular plants (Fischer et al. 2019).

3.2   Interviews on Health Issues

All interview partners described the health infrastructure and medical provision as good and sufficient for both municipalities (see the complete interviews in Appendix 2). At the same time, the way of life clearly has potential for a change towards a healthier one—reducing the need for medical interventions. Traditional medicines, schnapps, and fresh and healthy foods and herbs are all getting increasing attention in the municipalities both for personal use and as a touristic attraction.

4   Discussion

The data presented in this study relate to Goal 3 of the UN Sustainable Development Goals, "Ensure healthy lives and promote well-being for all at all ages", as well as SDG 15, "Protect, restore and promote sustainable use of terrestrial ecosystems, sustainably manage forests, combat desertification, and halt and reverse land degradation and halt biodiversity loss", but gaps in knowledge are evident. By combining data on changes in land cover and livelihoods in two Alpine villages, we try to relate sustainable development of the landscape to human health and well-being.

In Austria, one of the richest countries in the world, healthcare in the communities is very good, with low maternal mortality and an environment that sustains health. As evident from the interviews, lifestyle is considered a major criterion for health and well-being. Interviewees expected the most room for improvement at this individual level. To identify this potential, our land cover and demographic data need to be complemented by empirical social research. What is clear, however, is that only together are scientists and local people able to approach any problems in a broader and more effective way—and hopefully find answers.

Keeping in mind the historical background of the partial deforestation of Galtür in the course of timber production for the salt mines in Hall, present changes in land cover are less drastic and have minor effects. Even accounting for the increased area used for settlements and the skiing industry, every type of land use is still present. Changes in the landscape are found mainly in Ischgl during the summer, as the arable fields of the 1950s have been replaced by meadows—a general phenomenon in the Alps for the second half of the twentieth century. The forest is now regaining areas which had been used for grazing cattle, and warmer temperatures allow trees to conquer areas occupied by glaciers 100 years ago. With ongoing debates about land use at all levels of society and politics, awareness of land-use changes and their implications may be greater than ever in history.

5 Conclusion

In the communities of Ischgl and Galtür, land use/land cover has changed, with an increasing portion of previously used land being abandoned. Changes in climate have led to glacial recession with biotic succession and a rise in the treeline. Both factors have impacted biodiversity. Human impact on the landscape since 1857 comprises an increase in the settlement areas, a decrease in arable fields, as well as the addition of traffic and tourism infrastructure. The areas used for ski tourism are also part of alpine pastures.

Traditional health-related activities are the ongoing use of *Gentiana punctata* as a medicinal drug and as a herb, although for a short period, use of the roots of this protected plant was forbidden. Now, strictly limited use is allowed again and protected in the United Nations' framework for cultural heritage.

The local people of the communities of Ischgl and Galtür see the landscape they live in as an intrinsic part of their livelihood, and one they benefit from for their own health and for health tourism. Living standards in the middle of the European Alps are already very high in terms of health infrastructure, food and water security, affordable and clean energy, and climate change mitigation and adaptation. SDG indicators for Austria confirm the statements in the interviews that future optimisation of health and well-being does not need additional resources and infrastructure, but a change in the way of life. In the light of the "One Health" approach and the aim of "living in harmony with nature", the most important piece of the puzzle to be improved now is the individual lifestyles to increase personal well-being and resonance with nature.

Acknowledgements We thank the communities of Ischgl and Galtür as well as the Alpinarium, the Jamtal Environmental Education Centre, and the Jamtal hut for their cooperation and support. The Federal Administration of Tyrol provided Open Government geodata and statistics. Last but far from least we thank Brigitte Scott for improving the language!

Appendix 1

Table 8.5 Targets and indicators for SDG 3: Ensure healthy lives and promote well-being for all at all ages (Independent Group of Scientists appointed by the Secretary-General, 2019)

Aim	Indicator
3.1 By 2030, reduce the global maternal mortality ratio to less than 70 per 100,000 live births	3.1.1 Maternal mortality ratio
	3.1.2 Proportion of births attended by skilled health personnel
3.2 By 2030, end preventable deaths of newborns and children under 5 years of age, with all countries aiming to reduce neonatal mortality to at least as low as 12 per 1000 live births and under-5 mortality to at least as low as 25 per 1000 live births	3.2.1 Under-5 mortality rate
	3.2.2 Neonatal mortality rate
3.3 By 2030, end the epidemics of AIDS, tuberculosis, malaria, and neglected tropical diseases and combat hepatitis, waterborne diseases, and other communicable diseases	3.3.1 Number of new HIV infections per 1000 uninfected population, by sex, age, and key populations
	3.3.2 Tuberculosis incidence per 100,000 population
	3.3.3 Malaria incidence per 1000 population
	3.3.4 Hepatitis B incidence per 100,000 population
	3.3.5 Number of people requiring interventions against neglected tropical diseases
3.4 By 2030, reduce by one-third premature mortality from non-communicable diseases through prevention and treatment and promote mental health and well-being	3.4.1 Mortality rate attributed to cardiovascular disease, cancer, diabetes, or chronic respiratory disease
	3.4.2 Suicide mortality rate
3.5 Strengthen the prevention and treatment of substance abuse, including narcotic drug abuse and harmful use of alcohol	3.5.1 Coverage of treatment interventions (pharmacological, psychosocial, and rehabilitation and aftercare services) for substance-use disorders
	3.5.2 Alcohol per capita consumption (aged 15 years and older) within a calendar year in litres of pure alcohol
3.6 By 2020, halve the number of global deaths and injuries from road traffic accidents	3.6.1 Death rate due to road traffic injuries
3.7 By 2030, ensure universal access to sexual and reproductive healthcare services, including for family planning, information and education, and integration of reproductive health into national strategies and programmes	3.7.1 Proportion of women of reproductive age (aged 15–49 years) who have their need for family planning satisfied with modern methods
	3.7.2 Adolescent birth rate (aged 10–14 years; aged 15–19 years) per 1000 women in that age group
3.8 Achieve universal health coverage, including financial risk protection, access to quality essential healthcare services, and access to safe, effective, quality, and affordable essential medicines and vaccines for all	3.8.1 Coverage of essential health services
	3.8.2 Proportion of population with large household expenditures on health as a share of total household expenditure or income

(continued)

Table 8.5 (continued)

Aim	Indicator
3.9 By 2030, substantially reduce the number of deaths and illnesses from hazardous chemicals and air, water, and soil pollution and contamination	3.9.1 Mortality rate attributed to household and ambient air pollution
	3.9.2 Mortality rate attributed to unsafe water, unsafe sanitation, and lack of hygiene (exposure to unsafe water, sanitation, and hygiene for all (WASH) services)
	3.9.3 Mortality rate attributed to unintentional poisoning
3.a Strengthen the implementation of the World Health Organization framework convention on tobacco control in all countries, as appropriate	3.a.1 Age-standardised prevalence of current tobacco use among persons aged 15 years and older
3.b Support the research and development of vaccines and medicines for the communicable and non-communicable diseases that primarily affect developing countries, provide access to affordable essential medicines and vaccines, in accordance with the Doha Declaration on the TRIPS Agreement and Public Health, which affirms the right of developing countries to use to the full the provisions in the Agreement on Trade-Related Aspects of Intellectual Property Rights regarding flexibilities to protect public health, and, in particular, provide access to medicines for all	3.b Support the research and development of vaccines and medicines for the communicable and non-communicable
3.c Substantially increase health financing and the recruitment, development, training, and retention of the health workforce in developing countries, especially in least developed countries and small island developing states	3.c.1 Health worker density and distribution
3.d Strengthen the capacity of all countries, in particular developing countries, for early warning, risk reduction, and management of national and global health risks	3.d.1 International Health Regulations (IHR) capacity and health emergency preparedness

Appendix 2 Interview on Galtür and Health

Anton Mattle (AM), mayor of Galtür, and Helmut Pöll (HP), head of the municipal administration of Galtür, were interviewed in German on 2 February 2021, 10:10–10:30 a.m., by Andrea Fischer (AF). The transcript was translated into English.

AF: Question 1. Galtür and health: What is your assessment of the current healthcare provision?

HP: Healthcare provision is mainly covered by the local general practitioner. The nearest hospital is in Zams. It is well equipped and can be reached by car in about 40 min. Specialist doctors across the spectrum are available in Zams, Landeck, and Innsbruck. I would consider the medical provision very good.

Older people are being cared for either at home by members of their family or in the home for the elderly. There is support for care at home available from the local care services *Sozialsprengel*, for instance with looking after patients. The home for the elderly is run jointly with other municipalities and is located in the village of Grins at the start of the Paznaun Valley. Whether people are cared for at home or in the home for the elderly depends not just on their health status and the related medical and care needs, but also on the total circumstances of the family, for instance whether there is sufficient space available, whether all family members are working, or someone is at home.

AF: Question 2. Galtür and health: How is health in Galtür—how has the situation changed in recent decades?

HP: Health has declined. We no longer lead as healthy lives as the older people. For example, my father ate everything, including bacon with lots of fat, but then he did hours of physical labour. So he never had any problems with his weight, and he was never really ill and lived to over 90. So that fat never accumulated. Today we lack exercise. Of course, medicine has made great advances, and we can fix a lot. But our lifestyle is not as healthy as in the past.

AF: Question 3. What traditional substances and practices of healing exist in Galtür?

HP: Many people still use herbs in everyday life, for cooking and as tisanes. In our community we also have individuals who work with healing herbs as a sideline and produce oils and salves. We have also developed touristic options in this respect, for instance the learn-about-herbs path.

AM: In our village the spotted gentian [*Gentiana punctata*] is distilled into schnapps which is used as a traditional healing substance. Unlike the yellow gentian [*Gentiana lutea*], the spotted gentian cannot be cultivated. It is already mentioned way back in the herbarium of Admont monastery as a medical drug with a number of applications. Gentian of course also has a cult following. Marmot fat is another example, it is used against problems of the musculoskeletal system, but also in midwifery, for humans and animals.

AF: Question 4. What is your view of health tourism?

AM: We work with what nature offers us, our landscape. We want to preserve nature, our ancestors rejected plans to develop glacier skiing. Our village is part of the climate association and we participated in a climate change adaptation project. We are the only climatic spa in Tyrol and have been awarded the ECARF label of the European Centre for Allergy Research Foundation.

Appendix 3 Interview on Ischgl and Health

Christian Schmid (CS), Ischgl local authority manager, was interviewed on 4 February 2021, 13:00–13:30 by Andrea Fischer (AF)

AF: Question 1. What is the current health infrastructure like?

CS: General practitioner Dr. Walser employs four doctors. Then there is the Schenk sports injury surgery, equipped with a CT scanner; they also operate two rescue helicopters in the skiing area, but only in the winter season from late November to early May. We have two ambulances of the Red Cross stationed in the village, plus another rescue helicopter of the Austrian automobile club ÖAMTC stationed all year round in Finais/Zams. One dentist and two physiotherapists have their practices in the village. The home for the elderly is located in Grins, and the Ischgl care association supports the infrastructure for care at home.

AF: Question 2. In your view, has the health situation improved or declined in recent decades?

CS: In the past there was only one ambulance and one general practitioner in the village, so provision has improved. But the people practiced more exercise in the past, and possibly also more leisure time. It would be good to be free every day from about 4 p.m. and have the weekends free. There is a rethink among younger people; they roam our mountains again in summer and winter, getting exercise.

AF: Question 3. Is there traditional medicine in Ischgl?

CS: In Galtür there is the Enzner (schnapps) (laughs). In Ischgl we have good cuisine—ten gourmet restaurants and five-star chefs. Gathering gentian is not as popular here as in Galtür, but the restaurants in our village and up on Idalpe use local produce. There is a small slaughterhouse in Ischgl and a farm cheesemaker. A range of produce of the 60 farmers in our municipality is processed and consumed in our village.

AF: Question 4. Is there health tourism in Ischgl?

CS: In 1969 we created a forest swimming pool with tennis courts (pulled down and rebuilt in 1993). In summer you can walk, climb, cycle, and play tennis, and in winter you can ski, snowboard, go cross-country skiing, ice skate, and much more. In short, enjoy nature in its beauty to the full. The open-air swimming pool and the leisure centre, built in 1986, are beginning to show their age. A new spa is under construction.

References

Bender, O. (2015). Study on immigration to and emigration from the Alps with respect to the 'new highlanders", In Permanent Secretariat of the Alpine Convention (Ed.), *Demographic changes in the Alps. Report on the state of the Alps* (pp. 56–60), Alpine Signals Special Edition 5, Alpine Convention. Retrieved from www.alpconv.org/en/home/news-publications/publications-multi media/detail/rsa5-demographic-changes-in-the-alps/

Bender, O., & Borsdorf, A. (2014). Neue Bewohner in den Alpen? Räumliche Mobilität und Multilokalität in Tirol. In T. Chilla (Ed.), *Leben in den Alpen. Verstädterung, Entsiedlung und neue Aufwertungen. Festschrift für Werner Bätzing zum 65. Geburtstag.* Haupt, pp. 15–30.

Bender, O., & Haller, A. (2017). The cultural embeddedness of population mobility in the Alps: Consequences for sustainable development. *Norsk Geografisk Tidsskrift/Norwegian Journal of Geography, 71*(3), 132–145. https://doi.org/10.1080/00291951.2017.1317661

Bender, O., & Kanitscheider, S. (2012). New immigration into the European Alps: Emerging research issues. *Mountain Research and Development, 32*(2), 235–241. https://doi.org/10. 1659/MRD-JOURNAL-D-12-00030.1

Böhm, H. (1970). Das Paznauntal. *Die Bodennutzung eines alpinen Tales auf geländeklimatischer, agrarökologischer und sozialgeographischer Grundlage* (= Forschungen zur deutschen Landeskunde 190), Bundesforschungsanstalt für Landeskunde und Raumordnung. Bonn-Bad Godesberg.

Brugger, B., & Erschbamer, B. (2012). Die Bergwiesen der Pidigalm (Gsiesertal, Südtirol): Auswirkungen der Planierung, Düngung und Mahd auf die Artenvielfalt. *Gredleriana, 12*, 39–66.

Collins, S., Swinton, S., Anderson, C. W., Benson, B., Brunt, J., Gragson, T., Grimm, N., Grove, M., Henshaw, D., Knapp, A. K., Kofinas, G., Magnuson, J., McDowell, B., Melack, J., Moore, J., Ogden, L., Porter, J., Reichman, O. J., Robertson, G. P., Smith, M. D., Vande Castle, J., & Whitmer, A. (2007). *Integrative science for society and environment: A strategic research initiative*, viewed 30 July 2021. Retrieved from www.researchgate.net/publication/304047732_ Integrated_Science_for_Society_and_the_Environment_A_strategic_research_initiative.

Dietre, B., Walser, C., Lambers, K., Reitmaier, T., Hajdas, I., & Haas, J. N. (2014). Palaeoecological evidence for Mesolithic to medieval climatic change and anthropogenic impact on the Alpine flora and vegetation of the Silvretta Massif (Switzerland/Austria). *Quaternary International, 353*, 3–16. https://doi.org/10.1016/j.quaint.2014.05.001

Federal Administration of Tyrol. (2021). data.tirol.gv.at, viewed 10 February 2021. Retrieved from www.data.gv.at/katalog/dataset/3051AF8F-D7F7-4FCB-AE6F-D18EF493F6FA.

Fischer, A., Fickert, T., Schwaizer, G., Patzelt, G., & Groß, G. (2019). Vegetation dynamics in Alpine glacier forelands tackled from space. *Nature Scientific Reports, 9*, 13918. https://doi.org/ 10.1038/s41598-019-50273-2

Fischer, A., Schwaizer, G., Seiser, B., Helfricht, K., & Stocker-Waldhuber, M. (2021). High-resolution inventory to capture glacier disintegration in the Austrian Silvretta. *Cryosphere, 15*, 4637–4654. https://doi.org/10.5194/tc-15-4637-2021

GALPIS—Geographical Alpine Information System. (2021). Unpublished database of temporally harmonised Austrian municipality data 1971–2011. IGF.

Gemeinde Galtür (ed.) (1999). Galtür. Zwischen Romanen, Walsern und Tirolern.

Huhn, N. (1997). Galtür und Ardez. Geschichte einer spannungsreichen Partnerschaft. Doctoral thesis, University of Innsbruck.

Independent Group of Scientists appointed by the Secretary-General. (2019). *Global sustainable development report 2019: The future is now—Science for achieving sustainable development.* United Nations.

Kääb, A., Reynolds, J. M., & Haeberli, W. (2005). Glacier and permafrost hazards in high mountains. In U. M. Huber, H. K. M. Bugmann, & M. A. Reasoner (Eds.), *Global change and mountain regions* (Advances in global change research) (Vol. 23, pp. 225–234). Springer. https://doi.org/10.1007/1-4020-3508-X_23

Krainer, K., & Ribis, M. (2012). A rock glacier inventory of the Tyrolean Alps (Austria). *Austrian Journal of Earth Sciences, 105*(2), 32–47.

Kutschera, W., Patzelt, G., Wild, E. M., Haas-Jettmar, B., Kofler, W., Lippert, A., Oeggl, K., Pak, E., Priller, A., Steier, P., Wahlmüller-Oeggl, N., & Zanesco, A. (2014). Evidence for early human presence at high altitudes in the Ötztal Alps (Austria/Italy). *Radiocarbon, 56*(3), 923–947. https://doi.org/10.2458/56.17919

Land Tirol. (2021). *Tourism in Tyrol*, viewed 8 May 2021. Retrieved from www.tirol.gv.at/statistik-budget/statistik/tourismus/#c76985.

Lichtenberger, E. (1965). Das Bergbauernproblem in den österreichischen Alpen Perioden und Typen der Entsiedlung (The problem of mountain farming in the Austrian Alps). *Erdkunde, 19*(1): 39–57.

Lutz, W., Striessnig, E., Dimitrova, A., Ghislandi, S., Lijadi, A., Reiter, C., Spitzer, S., & Yildiz, D. (2021). Years of good life is a well-being indicator designed to serve research on sustainability. *Proceedings of the National Academy of Sciences, 118*(12), e1907351118.

Mayer, R., Kaufmann, R., Vorhauser, K., & Erschbamer, B. (2009). Effects of grazing exclusion on species composition in high-altitude grasslands of the Central Alps. *Basic and Applied Ecology, 10*(5), 447–455. https://doi.org/10.1016/j.baae.2008.10.004

Pauli, H., Gottfried, M., Dullinger, S., Abdaladze, O., Akhaltkatsi, M., Benito Alonso, J. L., Coldea, G., Dick, J., Erschbamer, B., Fernández Calzado, R., Ghosn, D., Holten, J., Kanka, R., Kazakis, G., Kollár, J., Larsson, P., Moiseev, D., Moiseev, P., Molau, U., . . . Grabherr, G. (2012). Recent plant diversity changes on Europe's mountain summits. *Science, 336*, 353–355. https://doi.org/10.1126/science.1219033

von Pfister, O. (1911). *Das Montavon mit dem oberen Paznaun. Ein Taschenbuch für Fremde und Einheimische*. Jungen & Sohn, 224 pp.

Scharr, K. (2018). Der Franziszeische Kataster und seine Rolle im Kaisertum Österreich (1817–1866). *Österreich in Geschichte und Literatur mit Geographie, 2*, 120–138.

Schmitt, M. (2010). Agritourism—From additional income to livelihood strategy and rural development. *The Open Social Science Journal, 7*(3), 41–50. https://doi.org/10.2174/1874945301003010041

Statistik Austria. (2021) viewed 10 February 2021. Retrieved from www.statistik.at/blickgem/gemDetail.do?gemnr=70608.

Ulmer, F. (1943). Die Schwabenkinder. *Ein Beitrag zur Sozial- und Wirtschaftsgeschichte des westtiroler Bergbauerngebietes*, Wissenschaft und Volk 1.

UNESCO. (2013). *Wissen um die Standorte, das Ernten und das Verarbeiten des punktierten Enzians*, viewed 10 February 2021. Retrieved from www.unesco.at/kultur/immaterielles-kulturerbe/oesterreichisches-verzeichnis/detail/article/wissen-um-die-standorte-das-ernten-und-das-verarbeiten-des-punktierten-enzians.

United Nations Statistical Commission. (2017). *Global indicator framework for the Sustainable Development Goals and targets of the 2030 Agenda for Sustainable Development*, adopted by the General Assembly in A/RES/71/313, viewed 23 July 2021. Retrieved from https://unstats.un.org/sdgs/indicators/Global%20Indicator%20Framework%20after%202021%20refinement_Eng.pdf, viewed on 23.07.2021; data base https://unstats.un.org/sdgs/indicators/database/.

Zwitter, Ž., & Rasran, L. (2022). *Species-rich grasslands in the Alps in the last millennium: Environmental history and historical ecology*. Austrian Academy of Sciences (forthcoming).

The opinions expressed in this chapter are those of the author(s) and do not necessarily reflect the views of UNU-IAS, its Board of Directors, or the countries they represent.

Open Access This chapter is licenced under the terms of the Creative Commons Attribution 3.0 IGO Licence (http://creativecommons.org/licenses/by/3.0/igo/), which permits use, sharing, adaptation, distribution and reproduction in any medium or format, as long as you give appropriate credit to UNU-IAS, provide a link to the Creative Commons licence and indicate if changes were made.

The use of the UNU-IAS name and logo, shall be subject to a separate written licence agreement between UNU-IAS and the user and is not authorised as part of this CC BY 3.0 IGO licence. Note that the link provided above includes additional terms and conditions of the licence.

The images or other third party material in this chapter are included in the chapter's Creative Commons licence, unless indicated otherwise in a credit line to the material. If material is not included in the chapter's Creative Commons licence and your intended use is not permitted by statutory regulation or exceeds the permitted use, you will need to obtain permission directly from the copyright holder.

Chapter 9
Promoting Local Health Traditions and Local Food Baskets: A Case Study from a Biocultural Hotspot of India

N. Anil Kumar, V. V. Sivan, and P. Vipindas

Abstract Degradation of socio-ecological production landscapes (SEPLs) triggered mainly by the impoverishment of biodiversity and the increasing incidence of climate catastrophes significantly challenges human health and food and nutritional security. Critical concern needs to be placed on ensuring both human and ecosystem health and contributing to nutrition-sensitive local food production and protection of SEPLs. As case points, we describe herein a few interventions and their impacts in promoting the *conservation*, *cultivation*, *consumption*, and *commercial* aspects regarding the medicinal and food plant diversity of a biocultural diversity hotspot in the Malabar region of India. The local communities of this region have historically possessed a wide array of local health traditions (LHTs) and local food baskets (LFBs) based on a landscape approach. Yet, this richness is being eroded or oversimplified, and as a result, many plants important for their local food and health value are becoming rare. The need for revitalisation of the LHTs and LFBs through homestead and landscape-level interventions is discussed in view of human immunity to infectious diseases. Recommendations are also suggested to address some of the policy gaps in promoting the sustainable management of SEPLs.

Keywords Biodiversity · Local health and food traditions · Self-help groups · Primary healthcare training · Home nutrition gardens · Immunity

1 Context

Biodiversity is the major link between environment, food, medicine, and human health. Although areas rich in biodiversity have a correspondingly high number of potential pathogens that can impact human health negatively, these viruses often live safely in animal reservoirs. The services of ecosystems that are rich in biodiversity

N. Anil Kumar (✉) · V. V. Sivan · P. Vipindas
M. S. Swaminathan Research Foundation, Meppadi, Kerala, India
e-mail: anil@mssrf.res.in

© The Author(s) 2022
M. Nishi et al. (eds.), *Biodiversity-Health-Sustainability Nexus in Socio-Ecological Production Landscapes and Seascapes (SEPLS)*, Satoyama Initiative Thematic Review, https://doi.org/10.1007/978-981-16-9893-4_9

often have the capacity to shield as a protective factor to prevent transmission of infectious diseases (Romanelli et al., 2015). The research and communication needs, however, remain enormous for our modern society to valuate such services of the ecosystems, for both wild and socio-ecological production landscapes (SEPLs), and to understand the interlinkages between human, environment, and animal health. Health traditions of pre-modern societies across the globe have treated environment and human health synergistically as one health to fight many human health problems. For instance, in the Indo-Malayan realm, the Indian region in particular has an age-old non-institutional practice of local health traditions (LHTs) and maintenance of local food baskets (LFBs), with the active participation of around one million traditional healers and around 200 million informed households involving the use of over 7500 plant species (Arima & Harilal, 2018). The holistic systems of medicine that originated in India, such as Ayurveda (an alternative medicine system which postulates "Food itself is medicine and medicine itself is food"), *Sidha* (a traditional Tamil system of medicine), yoga, naturopathy, and Sowa-Rigpa (enriched in the Trans-Himalayan region and has similarity with Ayurvedic philosophy), put equal emphasis on the mind, body, and spirit and strive to manage health in a holistic manner. Diverse methods for improving immunity in both codified and non-codified systems and ranging from herbal water and tea, *Choorna* (powdered form), *Lehyam* (colloid form), and *Kashaya* (tinctures) to complex rejuvenation therapies are known in India. Plants for these preparations are often harvested from the community landscapes (SEPLs). The community landscapes are in general maintained with a mosaic of vegetation types, such as community forests, sacred groves (remnants of past evergreen forests untouched due to religious beliefs), moist bamboo brakes, bushes, scrubs, riparian forests, swamps, and marshy grasslands.

Sen and Chakraborty (2017) reported local trade of more than 1500 herbals as dietary supplements or traditional ethnic medicines in India. These therapeutic and dietary domains together produce up to 25,000 effective plant-based formulations that are commonly used by a large majority of rural and ethnic people. Although India has rich plant species with diverse folk medicinal usages recorded in the community-managed landscapes, only a fraction of these species have been studied scientifically for their pharmacological potential or health benefits.

The case reported here from the Malabar region of India is intended to throw light on the need for increasing awareness on the availability of and informed access to local medicinal and food plant diversity, healthy diets, and a favourable policy environment related to primary healthcare and production and consumption of healthy foods. The chapter points towards the need for scientific understanding of the interlinkages between the ecological heterogeneity of production landscapes, traditional resource management practices, and abundance or scarcity of plants of therapeutic, food, and health value to local communities.

2 About the Intervention Site

The intervention took place in the Wayanad district, a hilly terrain in the historic and globally renowned maritime destination that is the Malabar region. Wayanad terrain, covering an area of 2172 sq. km, encompasses the southwest part of the Nilgiri mountain range ($76°15'$–$77°15'$E and $11°15'$–$12°15'$N). It is an east-sloping, gently undulating, medium-elevation plateau abruptly descending in the west to the Kerala plains but merging imperceptibly with the Mysore plateau to the east. Malabar is known for its rich diversity of spices like black pepper, ginger, turmeric, cardamom, and medicinal plants. For Malabar, the historic treatise of van Rheed's "*Hortus Malabaricus*", published between 1678 and 1693, is a testimony of its rich herbal heritage. Malabar is one of the richest centres of endemic, endangered biodiversity and cultural diversity of the country (Fig. 9.1). The basic information on the intervention site is provided in Table 9.1.

The biodiversity of Wayanad is greatly influenced by the management approaches of the diverse "sociocultural groups" that live there (Fig. 9.2). There are over 36 ethnic communities in Kerala, of which 11 are seen in the district. Farming communities in Wayanad cultivate more than 100 taxa of edible roots, tubers, and rhizomes; 20 varieties of legumes; 16 citrus cultivars; 150 vegetable varieties; and about 14 Musa cultivars (M. S. Swaminathan Research Foundation (MSSRF), 2020).

Wayanad's population counts 817,420 people (401,684 males and 415,736 females) with a tribal population of 148,215 members, which constitutes 18.1% of

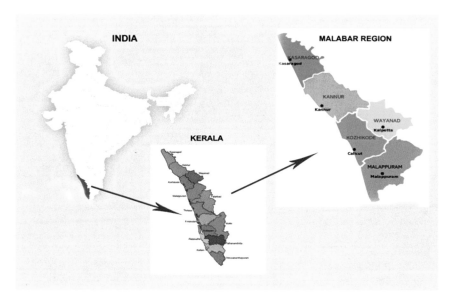

Fig. 9.1 Map of Kerala state and Malabar region (source: authors' creation based on map retrieved from Wikimedia Commons Contributors, 2021)

Table 9.1 Basic information of the study area (source: Government of Kerala, 2013)

Country	India
Province	Kerala
District	Wayanad
Municipality	n.a.
Size of geographical area (hectare)	213,100
Number of direct beneficiaries (persons)	51,320
Number of indirect beneficiaries (persons)	81,742
Dominant ethnicity(ies), if appropriate	Muslims; however, 18.53% of the population are Adivasis (indigenous communities)
Size of the case study/project area (hectare)	134,313
Geographic coordinates (latitude, longitude)	11.6854° N, 76.1320° E

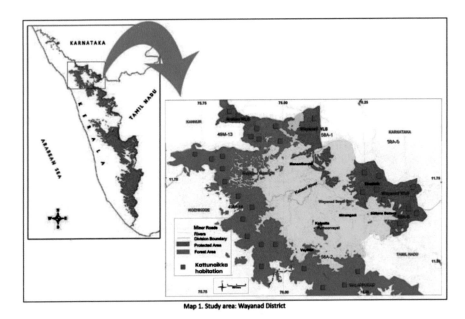

Map 1. Study area: Wayanad District

Fig. 9.2 Map of Kerala state and Wayanad district (source: authors' creation based on map retrieved from Wikimedia Commons Contributors, 2021)

the total population (Census of India, 2011). *Paniya* is the largest tribal group in Kerala, and 74.49% of the population of this tribe lives in this district. The *Katunaikka*, a forest-dependent community belonging to the Particularly Vulnerable Tribal Group (PVTG) category, also has a wide presence. Scores on basic human development indices for education and health are alarming among both these communities.

The terrain is a mosaic of diverse ecosystems ranging from mid-elevation plateau to high-range mountain landscapes with wet evergreen forests, and grasslands to dry-deciduous and scrub jungles. Apart from the wild vegetation types, the food and agricultural production landscapes are seen with certain edaphic types such as reed brakes, moist bamboo brakes, secondary evergreen forests, pseudo sholas, and marshy grasslands. Part of Wayanad's forests is a wildlife sanctuary protected under the Nilgiri Biosphere Reserve (NBR).

3 The Intervention

The intervention of M. S. Swaminathan Research Foundation (MSSRF) in the herb-based healthcare sector among the local communities, entitled the Green Health Programme (in partnership with the Department of Biotechnology, Government of India, and a local NGO called Vanamoolika), aimed at revitalisation of primary healthcare practices and was put into action during four intermittent time periods (1998 to 2001; 2004 to 2006; 2008 to 2010; and 2015 to 2018) with the following five major activities:

1. *Formation and capacity-building of women's self-help groups (SHGs) and farmers' forums* under the supervision of traditional healers, Ayurvedic physicians, botanists, and nutrition experts
2. *Collection, conservation, and cultivation of medicinal plants* focusing on an inventory of medicinal plants known to the community and documentation of associated traditional knowledge
3. *Cultivation, validation, value addition, and marketing of medicinal rice* called *Navara (Njavara/Nakara)*
4. *Collection, conservation, and consumption of wild food plants*
5. *Enriching the local food baskets by promoting home nutrition gardens*

4 Key Results

The results that emerged from the intervention revealed the SEPLs to be comprised of shade-grown coffee farms with diverse wild trees on hill slopes and integrated rice paddies in the valleys, alongside networked water canals and bushes, head ponds, and sacred groves, with many relict species still existing on hill tops and accessed by tribal and local men and women. How the interventions helped to improve human and environmental health is schematically shown in Fig. 9.3 and described herein.

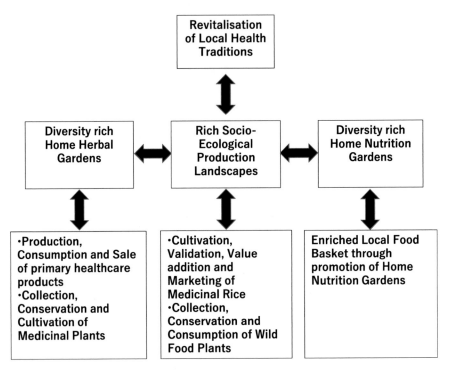

Fig. 9.3 Schematic illustration of the results of the interventions: The graphic shows how the interventions helped to improve human and environmental health. The revitalisation and popularisation of local health traditions (LHTs) and local food baskets (LFBs) have had impacts on both human and environmental health. Revitalisation and popularisation of LHTs helped to conserve plant diversity associated with LHTs and thereby plant diversity contributing to environmental health. Those who depend on LHTs benefit from good-quality primary healthcare with little economic cost (source: conceptualised by authors (N. Anil Kumar, V. V. Sivan, Vipindas P) for this case study to communicate various interconnections between human and environmental health)

4.1 Formation and Capacity-Building of Women's Self-Help Groups (SHGs) and Farmers' Forums

The Green Health Programme resulted in benefitting approximately 7000 rural and tribal women, including all the women SHGs and their family members and neighbours from Wayanad and the adjoining districts—Calicut and Malappuram. A total of 28 SHGs, each comprised of 10–15 women from resource-poor families, were trained, with group leaders serving as master trainers. Group members were encouraged to revitalise the practices and methods of the local healthcare systems, including medicinal plant identification, propagation, harvesting, preparation, storage, and use of primary healthcare products. Selected groups received 8 days of intensive training over a period of 8 months in the preparation of a set of 36 herbal formulations grouped into five categories: *choornam* (powder), *lehyam* (colloid form), *thailam* (oil), *gulika* (tablets), and powdered seed mixtures. One of the *lehyams* was "*chyavanaprash*", an ancient Ayurvedic product for boosting immunity prepared using 46 different ingredients (Table 9.2). Many of the medicinal herbs needed for this preparation were once commonly available in the traditionally managed agricultural landscapes.

The concept of "People's Biodiversity Registers" for recording of local knowledge was also shared with group members to facilitate the access and benefit-sharing mechanism. Not less than 500 awareness classes were conducted under this activity for SHG members. They were taken to observe various vegetation types, in particular the agricultural production landscapes, to help them understand the distributional differences of medicinal plants according to the land-use types. They were also allowed to visit well-reputed Ayurvedic institutions to get first-hand experience in their herbal gardens and drug production units, and to learn from their doctors.

The trained women were actively involved in the production and consumption of several primary healthcare products, household-level herbal preservation, and consumption of self-prepared products made from medicinal and wild edible plants. The trainees prepared these formulations for their own healthcare needs as well as to serve as an additional income source. The quality of the new formulations was assessed before marketing with the help of well-reputed research organisations. Nine of these products that were found to have high local demand were marketed by the SHGs. Marketing was conducted through neighbourhood sales and local horticultural and seed fairs. One of the best-selling formulations, a "*Navadhanya* mixture" (a mixture of nine types of grains), was tested and found to be nutritionally richer than many similar products available in the market as per the results of a nutritional analysis. All groups were trained on marketing techniques and developed their own marketing strategies, with income from sales always divided proportionately among members.

The initiative began at the grassroots level and was scaled up to establish a fully equipped and licensed factory (Fig. 9.4) in Wayanad in 2008 with financial support from the state government. It is being managed by one of the partnering organisations named Vanamoolika, along with the women SHG members.

Another highlight was a state government-supported project for production of ten selected herbal products. This project paved the way for organising a medicinal plant cultivators' society named JEEVANI to help small-scale farmers cultivate and market medicinal plants. The 165 farmers registered as members in this society were linked with the Kerala Industrial Infrastructure Development Corporation (KINFRA) for large-scale cultivation of medicinal plants.

Following project activities, people who never had the habit of keeping more than ten essential species of medicinal plants started to grow many more medicinal herbs at their households. According to project records, the number of species maintained in households and nearby areas ranged from 25 to 200, and people understood their potential uses. This practice reduced the pressure on wild habitats, ensuring the sustainability and health of the ecosystem.

Table 9.2 Ingredients in the *chyavanaprash lehyam* (an ancient Ayurvedic product for boosting immunity)

No.	Malayalam name	Binomial name	Common/ English name
1	Munja root	*Premna serratifolia* L.	Headache tree
2	Koovala root	*Aegle marmelos* (L.) Corrêa	Bel tree
3	Payyani root	*Oroxylum indicum* (L.) Benth.Ex Kurz	Broken bones tree
4	Kumizhu root	*Gmelina arborea* Roxb.	Beechwood tree
5	Paathiri root	*Stereospermum tetragonum* DC.	Yellow snake tree
6	Kurunthotti root	*Sida alnifolia* L.	Bala
7	Orila root	*Desmodium gangeticum* (L.) DC.	Sal leaved desmodium
8	Moovila root	*Pseudarthria viscida* (L.) Wight & Arn.	Sticky desmodium
9	Kaattuzhunnu root	*Vigna radiata* (L.) R. Wilczek	Wild black gram
10	Thippali root	*Piper longum* L.	Long pepper root
11	Vanthippali	*Piper chaba* Trel. & Yunck.	Bangla thippali fruit
12	Naikkurana root	*Mucuna pruriens* (L.) DC.	Cowhage, velvet bean
13	Cheruvazhuthana root	*Solanum violaceum* Ortega	Small brinjal
14	Cheruchunda	*Solanum indicum* L.	Indian night shade
15	Karakkadakasringi	*Pistacia chinensis* subsp. *integerrima* (J. L. Stewart ex Brandis) Rech. f.	Chinese pistache
16	Keezharnelli	*Phyllanthus amarus* L.	Feather foil plant
17	Munthiringa	*Vitis vinifera* L.	Raisins
18	Adapathiyankizhangu	*Holostemma ada-kodien* Schult.	Swallowwort
19	Pushkaramoolam	*Inula racemosa* Hook. f.	Orris root
20	Akil	*Dysoxylum malabaricum* Bedd. ex C.DC.	White cedar
21	Kadukkathodu	*Terminalia chebula* Retz.	Chebulic myrobalan
22	Amruthu	*Tinospora cordifolia* (Thunb.) Miers	Giloy
23	Kaattumuthira root	*Dolichos trilobus* L.	Wild horse gram
24	Jeevakam idavakam	*Malaxis rheedii* B. Heyne ex Wallace	Jeevak
25	Kacholam	*Kaempferia galanga* L.	Galanga, sand ginger
26	Muthanga	*Cyperus rotundus* L.	Nut grass
27	Thazhuthama	*Boerhavia diffusa* L.	Spreading hog weed

(continued)

Table 9.2 (continued)

No.	Malayalam name	Binomial name	Common/English name
28	Kaatupayar root	*Vigna vexillata* (L.) A. Rich	Wild green gram
29	Sathavarikkizhangu	*Asparagus racemosus* Willd.	Asparagus
30	Amukkuram	*Withania somnifera* (L.) Dunal	Winter cherry
31	Chittelam	*Heracleum rigens* Wall. ex DC	Heracleum
32	Chengazhineerkizhangu	*Kaempferia rotunda* L.	Indian crocus
33	Chandanam	*Santalum album* L.	Sandalwood
34	Palmuthakku	*Ipomoea mauritiana* Jacq.	Giant potato
35	Adalodakam	*Justicia adhatoda* L.	Malabar nut
36	Kakkathondi	*Capparis sepiaria* L.	Wild caper bush
37	Nellikka	*Phyllanthus emblica* L.	Indian gooseberry
38	Neyyu		Ghee
39	Then		Honey
40	Nallenna		Gingelly oil
41	Jaggery		Raw cane sugar
42	Koovanooru	*Maranta arundinacea* L.	Arrowroot
43	Thippali	*Piper longum* L.	Long pepper fruit
44	Nagappoovu	*Mesua thwaitesii* Planch. & Triana	Indian chestnut
45	Elam	*Elettaria cardamomum* (L.) Maton	Cardamom
46	Elavangam	*Cinnamomum malabatrum* (Burm.f.) J.Presl	Cinnamon

4.2   Collection, Conservation, and Cultivation of Medicinal Plants

The trained SHG women also started developing herbal and edible food gardens on the premises of their homes using plants that are in high demand for making the herbal and food preparations. Herbal nurseries were set up in these gardens to raise the locally important plants that are used to prepare the 36 herbal formulations, as well as to supply identified medicinal plants to households, to promote community-level medicinal plant gardens, and for commercial cultivation (Tables 9.3 and 9.4). During the period from 2000 to 2012, the project provided seedlings of 50 species of medicinal plants important for making home remedies to 158 SHG members to set up their home herbal gardens. Also during this period, about 100,000 seedlings of medicinal plant species were raised and maintained by the SHGs in two community nurseries. The help of trained SHG members was extended to M. S. Swaminathan Research Foundation (MSSRF) to get duplicate collections of the herbal plants, which are raised sustainably by collecting only necessary seeds from wild plants, for

Fig. 9.4 Factory established for the production of healthcare products (top left), equipment (top right and bottom left), products (bottom right) (photo credit: Sivan V V 2008)

the MSSRF's botanical garden. One successful outcome was the conservation of superior varieties of seven selected medicinal plants: *Aristolochia tagala*, *Celastrus paniculatus*, *Embelia ribes*, *Embelia tsjeriam-cottam*, *Gloriosa superba*, *Rauvolfia serpentina*, and *Saraca asoca*.

4.3   Cultivation, Validation, Value Addition, and Marketing of Medicinal Rice

Navara is reported to have multiple uses. It is a nutritious, balanced, and safe food for people of all ages (Gopinath et al., 2008). The key result in this area was promotion of cultivation, validation, value addition, and market development of speciality rice varieties such as *Navara* and *Chennellu* (medicinal) and *Gandhakasala* and *Mullanchanna* (scented). The process involved collection and documentation, genetic purification (to establish distinct types of cultivars, if any), clinical validation, and biochemical analysis of *Navara*, and studies on value addition for the other three varieties, with a view of taking the commodities developed to non-traditional markets as a medicinal or nutraceutical package. The results were (1) cultivation of special varieties of rice in an area of 23 ha, involving 76 new farmers constituting five farmer clusters; (2) genetic purification of *Navara* rice at the on-farm level; (3) six value-added products developed from *Navara* rice

Table 9.3 List of medicinal plants recommended for home herbal gardens (Kerala)

Malayalam name	Binomial name	Common/ English name	Remarks
Dashapushpa (ten herbs traditionally significant to the people of Kerala)			
Cheroola	*Aerva lanata*	Mountain knot grass	Ten flowers/plants used during festivals and in folk medicine preparations
Karuka	*Cynodon dactylon*	Bermuda grass	
Kayyonni	*Eclipta prostrata*	False daisy	
Mukkutti	*Biophytum reinwardtii*	Reinwardt's plant	
Muyalcheviyan	*Emilia sonchifolia*	Cupid's shaving brush	
Nilappana	*Curculigo orchioides*	Golden eye grass	
Poovamkurunthal	*Vernonia cinerea*	Purple fleabane	
Thiruthali	*Ipomea obscura*	Lesser glory	
Valliuzhinja	*Cardiospermum halicacabum*	Balloon plant	
Vishnukranthi	*Evolvulus alsinoides*	Dwarf morning glory	
Thrikadu (group of three spicy ingredients used in Ayurveda)			
Inchi	*Zingiber officinale*	Ginger	• Helps to clear excess *kapha* or mucus from the body. Supports respiratory function. Improves digestion. Helps in weight management
Kurumulaku	*Piper nigrum*	Black pepper	
Thippali	*Piper longum*	Long pepper	
Hruswapanchamoolam (group of five non-tree roots used in Ayurveda)			
Njerinjil	*Tribulus terrestris*	Bullhead	Five non-tree ingredients of *Dashamoolarishtam*, recommended for abdominal and gastrointestinal discomforts
Cheruvazhuthina	*Solanum violaceum*	Indian night shade	
Kandakaarichunda	*Solanum xanthocarpum*	Yellow berried night shade	
Orila	*Desmodium gangeticum*	Sal leaved desmodium	
Moovila	*Pseudarthria viscida*	Sticky desmodium	
Vallipanchamoolam (group of five vines used in Ayurveda)			
Paalmuthukku	*Ipomea mauritiana*	Giant potato	Five important vines used in Ayurveda in various medicinal preparations

(continued)

Table 9.3 (continued)

Naruneendi	*Hemidesmus indicus*	Indian Sarsaparilla	
Chittamruthu	*Tinospora cordifolia*	Giloy	
Maramanjal	*Coscinium fenestratum*	Yellow vine	
Aaduthodappaala	*Tylophora asthmatica*	Emetic swallow-wort	
Panchakolam (group of five spicy herbs used in Ayurveda)			
Thippali	*Piper longum*	Long pepper	Five important heat-generating ingredients used in herbal product formulation
Kaattuthippali	*Piper mullesua*	Wild long pepper	
Kaattumulaku	*Piper trioicum*	Wild pepper	
Chethikkoduveli	*Plumbago indica*	Indian leadwort	
Chukku	*Zingiber officinale*	Ginger	
Panchathikthakam (group of five bitter herbs used in Ayurveda)			
Chittamruthu	*Tinospora cordifolia*	Giloy	Five bitter plants used in Ayurveda
Veppu	*Azadirachta indica*	Neem	
Aadalodakam	*Justicia adhatoda*	Malabar nut	
Putharichunda	*Solanum anguivi*	African eggplant	
Padolam	*Trichosanthes lobata*	Wild snake gourd	
Thrikantakam (three spiny or caustic herbs used in Ayurveda)			
Cheruvazhuthina	*Solanum violaceum*	Night shade	A group of three thorny plants used in Ayurveda
Chethikkoduveli	*Plumbago indica*	Indian leadwort	
Vizhal	*Embelia ribes*	False black pepper	

put under trial marketing through a women's initiative; and (4) characterisation of *Navara* golden yellow and black varieties.

The results also included reporting for the first time the occurrence of four distinct subtypes of *Navara* rice, which included both awned and awnless black-glumed and yellow-glumed grains (Yasmin et al., 2006). Further, more farmers were mobilised at the cluster level for *Navara* cultivation, and the development and promotion of a few value-added products were attempted (Kumar et al., 2003). The pilot effort to clinically evaluate the efficacy of *Navara* rice (yellow awnless) in collaboration with the Institute of Applied Dermatology in Kasaragod called for profiling these varieties

Table 9.4 List of medicinal plants for community herbal gardens

Malayalam name	Binomial name	Common/English name	Remarks
Nalpamara (group of four barks used in treating skin-related ailments in Ayurveda)			
Arayal	*Ficus religiosa*	Sacred fig	Group of four fig trees, whose powdered bark is used for skin care
Athi	*Ficus racemosa*	Cluster fig	
Ithi	*Ficus microcarpa*	Chinese banyan	
Peral	*Ficus benghalensis*	Banyan tree	
Thriphala (group of three fruits helpful in weight loss)			
Kadukka	*Terminalia chebula*	Chebulic myrobalan	• A group of three fruits, helpful in weight loss, acts as a detoxifier, cures digestive issues, helps in fighting infections, and enhances immunity; beneficial in maintaining oral hygiene and eye health, and treating gastric ulcers and urinary tract infections
Nelli	*Phyllanthus emblica*	Indian gooseberry	
Thanni	*Terminalia bellirica*	Belleric myrobalan	
Thrigandha (three fragrant woods used in Ayurveda)			
Akil	*Dysoxylum malabaricum*	White cedar	Group of three fragrant trees frequently used in Ayurveda
Chandanam	*Santalum album*	Sandalwood	
Rakthachandanam	*Pterocarpus santalinus*	Red sandal	
Brihatpanchamoolam (five tree roots used in Ayurveda)			
Koovalam	*Aegle marmelos*	Bel tree	Five tree roots used for the preparation of *Dasamoolarishtam,* a fermented herbal tonic prepared using the root of ten medicinal plants (in the Sanskrit language dasa = 10, moola = root)
Munja	*Premna serratifolia*	Headache tree	
Palakappayyani	*Oroxylum indicum*	Broken bones tree	
Paathiri	*Stereospermum tetragonum*	Yellow snake tree	
Kumizhu	*Gmelina arborea*	Beechwood tree	
Panchapallavam (five tender leaves)			
Mavu	*Mangifera indica*	Mango	Five plants whose tender leaves are used in Ayurveda
Njaval	*Syzygium cumini*	Indian blackberry	
Mathalanarakam	*Punica granatum*	Pomegranate	
Vilaarmaram	*Limonia acidissima*	Wood apple	
Koovalam	*Aegle marmelos*	Bel tree	

Fig. 9.5 Women engaged in medicinal rice cultivation (left), seed material exhibited for dissemination and exchange in Wayanad Seed Fest 2018 (right) (photo credit: Vipindas P 2019)

for the presence of specific components responsible for the unique characteristics and defining of benefits for hemiplegia and other neuromuscular disorders (Guruprasad et al., 2014, p. 63).

These efforts to promote high-value rice helped to check the pace of paddy field conversion to alternate crops in selected cluster areas of Wayanad, which in turn helped to enhance employment opportunities for women (Fig. 9.5, left).

4.4 Collection, Conservation, and Consumption of Wild Food Plants

A study conducted during the period 2002–2005 revealed that the tribal groups of Wayanad have extensive knowledge regarding wild food and are using a wide array of plants and animals, with variations among different tribal groups (Fig. 9.6, left). There were 372 wild edibles utilised by different tribal communities, which include 102 species of leafy greens, 19 species of *Dioscorea*, 40 species of mushrooms, 5 species of crabs, 39 species of fishes, and 5 types of honey (Narayanan et al., 2004; Kumar & Narayanan, 2011).

The trained tribal women of the SHGs were very interested in the wild food plants as many of the wild roots, fruits, and leaves are of high nutritional and health value to them. At one time, tribal communities had access to boundless diversity—roots and tubers, fruits and seeds, small animals that were trapped, and fish and crab that were caught, constituting a major source of their nutrition. Another result of the project was the introduction and maintenance of live collections of wild foods (Fig. 9.5, right), like *Colocasia* and *Dioscorea* tubers, in home gardens. Balakrishnan et al. (2003) reported the existence of ten species of wild *Dioscorea* and seven lesser known varieties in the southern part of the Western Ghats. They reported several morphologically distinguishable species such as *D. pentaphylla*, *D. wallichii*, *D. hamiltonii*, and *D. belophylla*. Among the wild species of *Dioscorea*, *Nallanoora* (*D. pentaphylla* var. *pentaphylla*) is the most commonly consumed tuber, and *D.*

Fig. 9.6 Tribal woman engaged in wild leafy green collection (left), and tribal family engaged in wild *Dioscorea* collection (right) (photo credit: Sivan V V 2019)

hamiltonii and *D. oppositifolia* are the most frequently consumed (Fig. 9.6, right). Different tribes consume nine kinds of *Dioscorea* tubers, of which the most preferred are *Kavalakizhangu* (*D. oppositifolia*) and *Noorakizhangu/Nallanoora* (*D. pentaphylla* var. *pentaphylla*). They consider the *Noorakizhangu* and *Kavalakizhangu* to be rich in starch and fat, and the *Narakizhangu* (*D. wallichii*) to be rich in fibre.

4.5 Enriching the Local Food Baskets by Promoting Home Nutrition Gardens and Wild Food Plants

The training activities focused on the capacity-building of SHG members, Panchayath Raj Institutions (a system of rural local self-government in India elected by the local people for the management of local affairs), *Anganwadi* (a type of rural child care centre) workers, school teachers and students, and other community members on topics including sustainable crop cultivation practices, nutrition, agriculture-nutrition linkages, health and hygiene-related aspects, and more. Training modules included basic concept of home nutrition approach, and basic concept of a healthy and balanced diet, and culinary practices, including infant and young child feeding (IYCF) practices.

To serve as grassroots centres for nutrition intervention, nutrition gardens were established in *Anganwadi* centres. Community conservation plots for banana, pulses,

Fig. 9.7 Tribal children engaged in food collection (left), home nutrition garden maintained by tribal family at tribal hamlet Puthoorvayal, Wayanad district (right) (photo credit: Vipindas P.)

Fig. 9.8 Tribal man engaged in the management of community conservation plot (left), members of the community harvesting different varieties of tubers at Madamkunnu tribal hamlet, Wayanad district (right) (photo credit: Vipindas P)

legumes, and leafy greens were established and maintained at selected sites in the Wayanad, Thrissur, and Kannur districts. Participatory activities such as these have become even more relevant in the wake of climate change impacts on food crop species, especially in biodiversity and climate hotspots. In addition, nutritional assessment sample surveys (anthropometric and dietary) and medical camps were completed among the target groups to detect commonly existing health ailments due to nutritional deficiency.

A major result in this area was the promotion of home nutrition gardens (HNGs) for the purpose of enhancing food availability at the household level, particularly nutrient-dense foods through judicious integration of crops with animal husbandry, apiaries, fish tanks, etc. The HNG (Fig. 9.7, left and right) activities resulted in increasing (1) availability of a number of underutilised, nutrient-rich traditional and wild food crops (Fig. 9.8, left and right) and (2) availability of eggs, meat, and fish through appropriate interventions involving backyard poultry, small ruminants, and freshwater aquaculture among nearly 1000 tribal families. Also as part of home nutrition garden initiative, 30 crop varieties (pulses, roots and tubers, banana, leafy greens, and fruits) were promoted. This crop combination assures access to food almost year-round. Alongside the HNGs, efforts were taken to promote the integration of apiaries among 245 selected families scattered over the district by providing each of them with two honey boxes. The families harvested and marketed about 1200 kg of wild honey. Honey boxes placed in the vicinity of tribal hamlets helped tribal women to harvest wild honey, an activity that was earlier practised only by men. The tribal department of the state government has selected this initiative as a model for their interventions in poverty reduction in tribal areas of Kerala.

Highly motivated and progressive men and women farmers from the target villages were inducted as community resource persons (CRPs). They were skilled in various aspects of conservation, cultivation, and consumption of locale specific food crops. The pre-intervention nutritional assessment surveys helped in gathering the target groups' food preferences in order to identify and promote suitable crops in the HNGs. Trainings were conducted targeting families on topics including crop management; maintenance of HNGs; nutritional value of crops promoted in the gardens; nutritional requirements of different genders and age groups; awareness on water, sanitation, and hygiene; and role and function of different nutrients.

Outcomes of the activities also included (1) an increased knowledge base on the region's forest food biodiversity; (2) conservation of forest food diversity including wild relatives of some of the crops and homestead of tribal hamlets; and (3) nutritional profiling of some of the selected wild food species.

5 Discussion

Communities derive multiple benefits from SEPLs. Some of the key benefits in the case of this programme were (1) increased access to safe, traditional, and diverse foods through homestead crop diversification; (2) revitalisation of some traditional healthcare practices, which helped people to address many simple ailments on their own and in a timely fashion without rushing to the hospital or to see medical practitioners, which was otherwise common; (3) home herbal gardens with at least 25 medicinal plants at every household; (4) increased involvement in the management of medicinal and nutrition gardens; (5) increased awareness on the conservation of surrounding biodiversity; (6) considerable reduction of expenditure for food and healthcare; and (vii) additional income from sales of healthcare products.

But, despite these benefits, there has been a steady erosion of cultural knowledge, including that related to the LHTs associated with biodiversity utilisation across the targeted locations. This is mainly due to changes in the societal values and behaviours, as well as changes in the way healthcare, food production, and consumption practices are promoted by the state.

There have been a few trade-offs of the benefits community derive from SEPLs. For instance, protecting agro-biodiversity, particularly the genetic diversity of fruits and roots, from crop raiding by wild animals that are legally protected under forest and wildlife acts, is challenging. Crop raiding by monkeys, birds, wild boars, and elephants has now become more frequent in many locations that are closer to forests. Another situation is the growing trend to disband traditional joint family systems to become independent small families in order to get maximum benefits from government-sponsored schemes.

The effective management of the SEPL is mutually beneficial and could be well measured through quantitative and qualitative assessments. These measurements include changes in health and nutrition status through periodical assessments, analysis of the dietary diversity supplements provided, and intensity and frequency of health issues. For the landscape, monitoring of diversified home herbal gardens and home nutrition gardens kept enriched with plants from the wilderness, changes observed in harvesting resources from the wild, use of diverse herbal plants and wild foods that supplement the health system, regenerative agricultural fields with availability of and access to fresh water and healthy soil, and government-sponsored programmes to promote tribal agriculture are some of the methods and indicators used to measure and understand effectiveness.

The SEPL in Wayanad is experiencing rapid land-use and land cover changes that negatively impact the availability of many wild species of food and medicinal value. For instance, despite the richness in the genetic diversity of medicinal and nutrient-dense rice in Wayanad that was developed and conserved mainly by local farmers, the low productivity and low profitability of these varieties alongside economic pressures are compelling the farmers to convert the rice paddies into banana plantations. During 2006–2015, the area under rice cultivation in Wayanad drastically decreased from 21,770 ha to 8000 ha (Government of Kerala (GoK), 2015). This shift from paddy production to banana farming has had a cascading effect on the paddy wetland ecosystem. Firstly, the trend has impacted rainwater storage and aquifer recharging, and secondly, it has impacted the livelihood options of tribal communities like the *Paniya*, who largely depend on this ecosystem for food, employment, and income. Finally impacts have been wielded on the environment, paving the way to dump larger quantities of highly toxic pesticides for banana cultivation.

Similarly, the once species-rich shade-grown coffee farms in the region have now experienced a shift towards monocropping and high-input farming. There have been deliberate attempts to grow coffee completely exposed to sun with the high input of agrochemicals aimed at maximising crop productivity and income. This change in farming practices and intensified use of farm inputs have become major sustainability issues, particularly in the context of climate vulnerabilities. The challenges associated with climate variations result in staggering poverty, mainly nutrition and health issues, especially among the resource-poor farming communities, and the tribal communities like *Paniya, Adiya,* and *Kattunaikka* in Wayanad. Nevertheless, the growing awareness of local people on the conservation and sustainable management of SEPLs in the present scenario of food adulteration and ecosystem degradation is a green signal and opportunity to move forward.

This demands focused attention on the preservation and restoration of traditionally managed SEPLs, biocultural diversity, and revitalisation of the herbal healthcare heritage backed by scientific studies, behavioural changes, and capacity development programmes for the promotion of local biodiversity and LHTs. To address these issues, we suggest the following six integrated strategic interventions:

First, a national-level commitment to revitalise location-specific, culturally significant, and time-tested healthcare and sustainable food production practices: Integrating local health traditions and traditional medicine in the sustainable development and health agenda of the world can benefit both biodiversity and human health. A report by the WHO and CBD (Romanelli et al., 2015) discussed in detail the criticality of developing and providing cost-effective, safe, and quality-proven traditional medicine and the involvement of traditional healers in building up healthcare systems with new knowledge. Adequate investment and effective integration between different departments that deal with food and nutrition and primary healthcare promotion are required to improve human and ecosystem health.

Second, reorientation of primary healthcare services from an exogenous approach to an endogenous approach, where the focus is on the promotion of healthy foods and traditional medicine. The WHO recognises traditional medicine as an integral part of LHT, and as the "total of the knowledge, skill, and practices based on the theories, beliefs, and experiences indigenous to different cultures, whether explicable or not, used in the maintenance of health as well as in the prevention, diagnosis, improvement, or treatment of physical and mental illness" (WHO, 2013). Approaches should treat human-animal-environmental health synergistically as "One Health" to fight against any human health problems.

Third, promotion of "Green Health" practices in every home: Promotion must involve knowledgeable individuals, trained primary health practitioners, custodian farmers, tribal and rural youth, and women to focus on micronutrient-rich (naturally or chemically biofortified plants) and immunity-boosting herbals in home gardens. This can be part of promoting the concept of local health heritage villages or biodiversity heritage sites as proposed in the National Biodiversity Act (2002), wherein biodiversity is mainstreamed in the production of food, nutrition, healthcare, and livelihood needs.

Fourth, sustainable value-chain development for certified primary health-boosting foods, products, and nutraceuticals by organising a cadre of trained women's groups in the form of herbal producer organisations: Sustainability in the supply chain occurs when management practices and governance impact positively on the socio-economic and environmental outcomes throughout the life cycles of goods and services. Sustainable commerce can be better illustrated through the 4C continuum promoted by MSSRF (M. S. Swaminathan Research Foundation (MSSRF), 2019). In this approach, *commerce* is done with the primary aim of creating an economic stake through options in livelihood security on small and marginal farmers, and promoted through a *consumption* pattern that enhances biodiverse diets, ensures food and nutrition security, and drives *cultivation* practices based on principles of organic farming. All of the three sector activities lead to *conservation* of biodiversity in situ, on-farm, and ex situ at off-farm levels. With this approach, communities can continue to engage in promoting conservation, cultivation, consumption, and local-level commercialisation of herbals, as well as plant genetic resources (PGRs) of food, nutrition, and therapeutic value.

Fifth, promotion of integrated scientific research in nutrition, health, and well-being to achieve health innovations and nutritional products from traditional and ethnomedicine systems through a transdisciplinary approach (by involving traditional healthcare practitioners and physicians of multiple medical sciences) to understand the bidirectional linkage between undernutrition and immunity issues and the role of nutrition and herbal remedies for a healthy life: Research in labs needs to focus on the effects of multi-micronutrients on immunity and for medical health conditions.

Sixth, restoration and preservation of SEPLs to protect from climate risks and enhance food security, water supply, and biodiversity: The buffer zones of the Wayanad district qualify as the most characteristic locations for ecosystem restoration in the Western Ghats, which itself is the highest human-populated biodiversity hotspot in the world. As 2021–2030 is the UN Decade on Ecosystem Restoration, there should be efforts from the state for a massive scale-up in the restoration efforts. This warrants a comprehensive Training of Trainers (ToT) programme in SEPLs management to offer to rural youth and officials in hill area land management, watershed development, agriculture, and allied sectors.

6   Conclusion

The COVID-19 pandemic crisis is an opportune time to seriously reflect on our indigenous traditional knowledge systems that have time-tested sustainable practices and principles in biodiversity management, and also prudence in maintaining the heterogeneity of food and agriculture production landscapes. Such practices and age-old values help local communities to keep their options alive for accessing a diverse spectrum of species of food and health value from the wild. In this context, the primary healthcare system planners need to develop a community-centric and holistic healthcare approach to effectively cater to the needs of the rural and tribal populations. The "Health for All" goal set by the United Nations in 2000 was to be achieved by adopting preventive measures, immunisation, and acquiring the capacity to access treatment and good nutrition at the individual level. Equally significant is integrating the preservation and protection of wild landscapes, ecosystems, and species diversity to succeed in this goal. This is possible if there is a transformation in the healthcare system approach by making the traditional healthcare sector as the fourth tier of the national healthcare system.

References

Arima, M., & Harilal, M. (2018). The making of local health traditions in India: Revitalisation or marginalisation? *Economic and Political Weekly, 53*(30), 41–50.

Balakrishnan, V., Narayanan, R. M. K., & Kumar, A. N. (2003). Ethnotaxonomy of Dioscorea among the Kattunaikka people of Wayanad District, Kerala, India. *Plant Genetic Resources Newsletter*, no. 135, pp 24–32, viewed 28 July 2021. Retrieved from https://www.bioversityinternational.org/fileadmin/PGR/article-issue_135-art_4-lang_fr.html.

Census of India. (2011). District census handbook Wayanad, *Village and Town wise Primary Census* Abstract, Census of India, Trivandrum.

Government of Kerala. (2013). Panchayath level statistics 2011, Department of Economics and statistics, Kerala, viewed on 21 December 2021. Retrieved from http://www.ecostat.kerala.gov. in/images/pdf/publications/Panchayat_Level_Statistics/data/2011/pswayanad2011.pdf.

Government of Kerala (GoK). (2015). Kerala economic review, State Planning Board, Government of Kerala, Thiruvananthapuram.

Gopinath, D., Singh, V., & Naidu, A. (2008). Nutrient composition and physiochemical properties of Indian medicinal rice—Njavara. *Food Chemistry, 106*(1), 165–171. https://doi.org/10.1016/j. foodchem.2007.05.062

Guruprasad, A. M., Saravu, N. R., Shanthakumari, V., Sushma, K. V., Kumar, A. N., & Parameswaran, P. (2014). *Navarakizhi* and *Pinda Sweda* as muscle-nourishing Ayurveda procedures in hemiplegia: Double-blind randomized comparative pilot clinical trial. *The Journal of Alternative and Complementary Medicine, 20*(1), 57–64.

Kumar, A. N., Gopi, G., & Balakrishnan, V. (2003). Medicinal rice of Kerala: Navara, a rice variety cultivated only in Kerala, is widely use in local healthcare systems as well as in Ayurveda for many ailments. *Amruthu, FRLHT, 2,* 14–16.

Kumar, A. N., & Narayanan, R. M. K. (2011). Diversity, use pattern and management of wild food plants of Western Ghats: A study from Wayanad district, viewed 12 February 2021. Retrieved from https://mssbg.mssrf.org/wp-content/uploads/2020/07/wild-food-plants-western-Ghats-website.pdf.

M. S. Swaminathan Research Foundation (MSSRF). (2019). *From 30 years to 2030: Achieving sustainable development goals and strengthening science for climate resilience.* M. S. Swaminathan Research Foundation.

M. S. Swaminathan Research Foundation (MSSRF). (2020). *Draft report conservation.* M. S. Swaminathan Research Foundation, Community Agrobiodiversity Center.

Narayanan, R. M. K., Swapna, M. P., & Kumar, A. N. (2004). *Gender dimensions of wild food management in Wayanad, Kerala.* Community Agrobiodiversity Centre & Utara Devi Resource Centre for gender and development. M. S. Swaminathan Research Foundation, MSSRF/RR/04/ 12.

Romanelli, C., Cooper, D., Campbell-Lendrum, D., Maiero, M., Karesh, W. B., Hunter, D., & Golden, C. D. (2015). *Connecting global priorities: Biodiversity and human health: A state of knowledge review.* World Health Organization and Secretariat of the Convention on Biological Diversity.

Sen, S., & Chakraborty, R. (2017). Revival, modernization and integration of Indian traditional herbal medicine in clinical practice: Importance, challenges and future. *Journal of Traditional and Complementary Medicine, 7*(2), 234–244.

WHO. (2013). *WHO traditional medicine strategy: 2014–2023*, WHO Geneva, viewed 13 October 2021. Retrieved from https://www.who.int/publications/i/item/9789241506096.

Wikimedia Commons Contributors. (2021). File:Wayanad Locator map.svg. *Wikimedia Commons, the free media repository*, viewed 12 February 2021. Retrieved from https://commons. wikimedia.org/w/index.php?title=File:Wayanad_Locator_map.svg&oldid=581468204.

Yasmin, S. H., Gipson, M., Nandakumar, P. M., & Kumar, A. N. (2006). Scope of Navara cultivation. In *Proceedings of the gene seminar* (pp. 3–6), Community Agrobiodiversity Center.

The opinions expressed in this chapter are those of the author(s) and do not necessarily reflect the views of UNU-IAS, its Board of Directors, or the countries they represent.

Open Access This chapter is licenced under the terms of the Creative Commons Attribution 3.0 IGO Licence (http://creativecommons.org/licenses/by/3.0/igo/), which permits use, sharing, adaptation, distribution and reproduction in any medium or format, as long as you give appropriate credit to UNU-IAS, provide a link to the Creative Commons licence and indicate if changes were made.

The use of the UNU-IAS name and logo, shall be subject to a separate written licence agreement between UNU-IAS and the user and is not authorised as part of this CC BY 3.0 IGO licence. Note that the link provided above includes additional terms and conditions of the licence.

The images or other third party material in this chapter are included in the chapter's Creative Commons licence, unless indicated otherwise in a credit line to the material. If material is not included in the chapter's Creative Commons licence and your intended use is not permitted by statutory regulation or exceeds the permitted use, you will need to obtain permission directly from the copyright holder.

Chapter 10
Safeguarding the Biodiversity Associated with Local Foodways in Traditionally Managed Socio-Ecological Production Landscapes in Kenya

Patrick Maundu and Yasuyuki Morimoto

Abstract Traditionally managed socio-ecological production landscapes (SEPLs) provide communities with a range of goods and services vital for livelihoods, including nutrition and health. In Kenya, many of these landscapes, encompassing the resources therein and the indigenous knowledge vital for optimising their value, are now under threat.

Utilising diverse traditional foods for the benefit of local communities has often been hampered by insufficient knowledge about the foods and negative attitudes towards them. For over two decades, a team from the National Museums of Kenya, Bioversity International, and their partners has been working with local communities to find out how local food resources can contribute more to community livelihoods, especially with regard to nutrition, health, and income. Understanding local food systems is a vital step. The team developed a methodology for involving local communities, and the youth in particular, to inventory their foodways. The documentation opened opportunities for research and development interventions. This chapter highlights three development and conservation case studies founded on foodways documentation: (1) promoting African leafy vegetables in Kenya; (2) safeguarding *kitete* (bottle gourd) by Kyanika Women's Group in Kitui, Kenya; and (3) utilising digital technology to educate consumers about healthy eating using local foods.

All cases have shown that converting underutilised local foods into main sources of nutrition and income opportunities, as well as conserving these foods in their environment, requires foodways documentation, community participation, and multi-stakeholder and multidisciplinary collaboration. Awareness on the nutritional

P. Maundu (✉)
KENRIK, National Museums of Kenya, Nairobi, Kenya

Alliance of Bioversity International and CIAT, Nairobi, Kenya

Y. Morimoto
Alliance of Bioversity International and CIAT, Nairobi, Kenya

© The Author(s) 2022

M. Nishi et al. (eds.), *Biodiversity-Health-Sustainability Nexus in Socio-Ecological Production Landscapes and Seascapes (SEPLS)*, Satoyama Initiative Thematic Review, https://doi.org/10.1007/978-981-16-9893-4_10

and health benefits of local foods was a key incentive for their conservation and a catalyst for the change in attitudes and eating habits.

Keywords Socio-ecological production landscapes · Foodways documentation · Underutilised local foods · Conservation · Human nutrition and health · Community involvement · Multi-stakeholder, Multidisciplinary interventions · Kenya

1   Introduction

In Kenyan rural community food systems, a huge diversity of food species, especially fruits and vegetables, are obtained from the immediate environment or socio-ecological production landscapes (SEPLs), where they may be cultivated, semi-wild (cultivated and also wild), or gathered from the wild. The high diversity is found at landscape, food group, species, and below-species levels. In addition, a rich food cultural heritage—including local knowledge of food production and utilisation, beliefs, oral traditions, ceremonies, and other practices—exists as part of these food systems and constitutes the foodways of the communities. Food is therefore not only for nutrition, but also for the well-being of the community as a whole, and thus is inextricably linked with the landscape.

A high diversity of underutilised traditional foods can be seen in African leafy vegetables (Maundu et al., 2009). Foodways documentation by the National Museums of Kenya (NMK) and Bioversity International in collaboration with local community groups recorded about 220 vegetable species used by various communities in Kenya (Maundu et al. 1999a, b; Kilifi Udamaduni Conservation Group, 2010). Of these, 90.9% were gathered from the local wild landscape (Fig. 10.1). In spite of this native diversity, a few exotic vegetable species have dominated the Kenyan market and kitchen in the recent past.

Indigenous food systems in Kenya have been undergoing continuous disruptions since the colonial era. Traditional foods were looked down upon as foods of the poor and less modernised. This attitude and the associated stigma led to neglect by researchers and development workers. Consequently, the use of these foods

Fig. 10.1 A chart showing the proportion of cultivated and wild leafy vegetable species for the 220 species recorded in Kenya (data source: Maundu, Njiro, et al., 1999b; Kilifi Udamaduni Conservation Group, 2010)

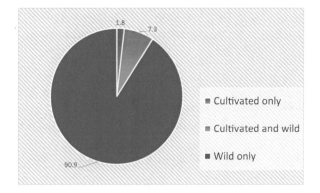

decreased, leading to erosion of local food diversity and the associated local knowledge.

By the year 2000, the diversity of leafy vegetables in the market and diet had been narrowed considerably, typically to three vegetables—cabbage, kale (*sukumawiki*), and Swiss chard ("spinach"). Traditional vegetables were mainly sold in very rural towns and backstreets of main cities. Fast "junk" foods such as potato fries were slowly replacing the healthier traditional cuisine, and younger people were increasingly inheriting less local knowledge from their communities. Such trends had the potential to lead to loss of local crop varieties, less dietary diversity, and hence food and nutrition insecurity (Maundu et al., 2008; Johns et al., 2013; Kariuki et al., 2013; Induli, 2019a).

1.1 Changing Diets and the Double Burden of Disease

Kenya, like many other African countries, now has the simultaneous problems of undernutrition (e.g. stunting in children) and over-nutrition. Stunting in children under 5 years in 2008–2012 was at 35.3%, reducing to 26% by 2014 (from Kenya Demographic and Health Survey 2014 in KNBS , 2014). On the other hand, lifestyle diseases such as high blood pressure, heart disease, and diabetes were on the rise. The prevalence of overweight and obesity among women of reproductive age (15–49 years) was 33% in 2014 (Kenya National Bureau of Statistics (KNBS), 2014) and on the rise. Rapid changes in dietary habits including high intake of carbohydrates, lipid-rich diets, low intake of micronutrients, and insufficient exercise are considered to be the main contributing factors. The urban population was most affected mainly due to overreliance on relatively cheaper, high-energy refined foods. The 2014 survey established that 28.6% of women from the urban population were overweight, and 14.9% were obese, whereas among women from the rural population, 18.9% were overweight and 7.1% were obese (Mkuu et al., 2018; Kenya National Bureau of Statistics (KNBS), 2014).

1.2 Interventions by a Multi-stakeholder, Multidisciplinary Team

In an effort to arrest the trends mentioned here, a multidisciplinary team from the National Museums of Kenya, Bioversity International, and partner institutions has been for the last two decades implementing a series of activities aimed at achieving the following: (1) safeguarding local food systems, (2) creating opportunities for community benefits, and (3) creating awareness about the value of local foods. Fundamental in this process is the documentation of the foodways of communities. With understanding of foodways, various development interventions based on opportunities offered by local food resources can be implemented (Induli et al., 2020). The following case studies of development initiatives are highlighted in this chapter: (1) African leafy vegetable (ALV) research and promotion, (2) bottle gourd (*kitete*) conservation, and (3) using modern technology to educate consumers on the nutritional and health value of local foods. A common objective of these initiatives has been finding incentives for communities to utilise these local food resources to improve their livelihoods while sustaining the landscapes from where these foods are obtained or grown. Understanding the benefits of these food resources, including nutrition and health, and also discovering new value through research collaboration, can motivate the communities to promote their local foods (Induli et al., 2020; JAICAF, 2020).

The objectives of this chapter therefore are firstly to demonstrate via these case studies the role of foodways documentation as a basis for developing interventions for biodiversity conservation and development at the community or landscape level, and secondly to demonstrate the importance of community participation and multi-stakeholder cooperation in finding ways to optimise the value of such resources as an incentive for conservation. Also highlighted in the chapter are the conditions necessary for the success of such interventions, key lessons learned, and shortcomings.

2   Activities (Methods)

2.1   The Foodways Documentation Method

In this section, we start on the premise that among Kenyan rural communities, food is diverse and not only cultivated, but also picked from wild environments, and that local knowledge is a prerequisite to optimising benefits gained from these resources. Much of this knowledge on food diversity is undocumented, making food-based and conservation interventions hard to implement. A logical starting point, therefore, would be to find answers to these questions: What foods are cultivated or gathered, from where, when, and by whom? How is the food source managed? How is the food harvested and utilised? Is it stored or sold? How is it prepared and eaten and who eats it? What do people say about its value to the body? Is it used in ceremonies or rituals and what are related beliefs or taboos? These are important questions in foodways documentation.

This approach is unique in two aspects. Firstly, it captures the totality of knowledge and practices related to food within a given community, and therefore results in a nearly complete inventory of food species used as well as ways of utilisation. Secondly, the documentation is executed by community members. Such documentation helps us understand the important place of the local landscape in people's food systems and resilience for food security and nutrition (Hama-Ba et al., 2016, 2017). In Kenya, traditional foodways are being eroded at a high rate due to global trends such as globalisation of food and urbanisation, which are not only detaching the youth from rural landscapes and from sources of local knowledge, but also changing

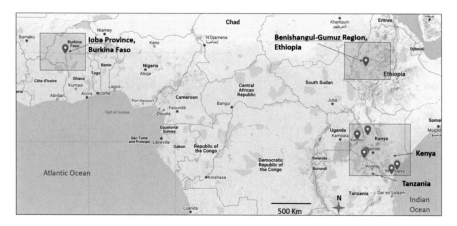

Fig. 10.2 Map showing countries and regions where the foodways method of documentation has been applied (source: Map Data © 2020 Google)

their preferences and values concerning food. Building the capacity of the youth, alongside that of other community members, to take the lead in foodways documentation helps bridge the intergenerational gap in the knowledge of local resources, a key ingredient in the sustainability of conservation interventions. Documentation provides the basic information necessary for other local resource-based intervention initiatives and also gives us the opportunity to recreate the foodways in the future in case the knowledge in the community is lost.

For a number of years, the authors have been developing a foodways documentation method that is community-led and involves the youth. The method has so far been tested in six communities across four African countries—the Mijikenda, Isukha, and the Pokot of Kenya; Loita Maasai of Kenya and Tanzania; Gumuz of Ethiopia; and Dagara of Burkina Faso (Figs. 10.2, 10.3, 10.4, and 10.5). Some basic information about these communities is presented in Tables 10.1, 10.2, and 10.3.

Documentation by Community Members

Selected community members are trained on foodways documentation. The basic materials needed for documentation include a notebook, a simple camera with video facility, or a smartphone and storage facility like a computer. A community coordinator or champion monitors the documentation process, gathers the collected information, and keeps it at a community resource centre or an institution within the community. This was the system applied among the Gumuz and Dagara. It is recommended that documentation goes on for a period of one year or more to capture variation through different seasons.

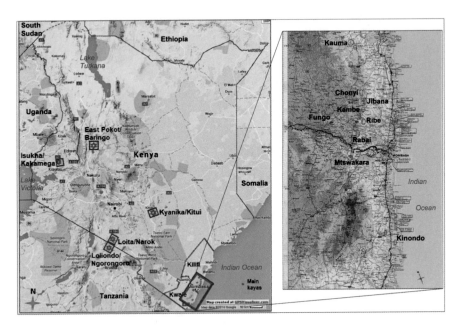

Fig. 10.3 Map showing the location of study sites in Kenya and northern Tanzania. The nine main sacred forests (*Kaya*) in coastal Kenya are shown in the inset (source: Map data © 2016 Google; source of inset base map: Survey of Kenya, 1973)

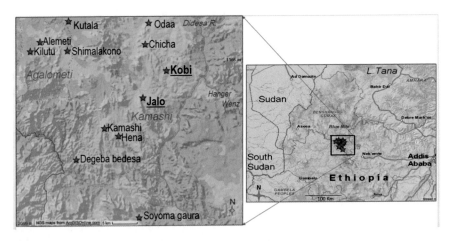

Fig. 10.4 Map of Kamashi area of Benishangul-Gumuz region in northwestern Ethiopia, showing the location of research villages (source of base map: ArcGISOnline.com, 2015; source of inset: ESRI/ArcGIS, 2015)

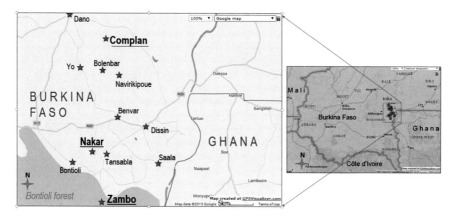

Fig. 10.5 Map of southwestern Burkina Faso, showing the location of research villages, with main villages underlined (sources: Base map: Map data © 2015 Google; inset map: ArcGISOnline.com, 2015)

Foodways Documentation by Schools

A most effective approach is mobilising and training the youth, including school children, to gather information from their parents and relatives and sharing this, for instance, using social media or with fellow pupils and their teachers. Documentation by school pupils has been done in four communities—the Isukha, Pokot, Mijikenda, and Maasai. Except for the Mijikenda, where nine schools were involved, two schools participated per community (Induli, 2019b).

In each school, two or more teachers volunteer to guide the pupils. Teachers could also be patrons of a school club. One teacher coordinates all the activities at a school. The teachers are trained by a multidisciplinary team in information gathering, including interviewing, organising the stories, how to use a camera, and handling data from pupils (e.g. preparing photo metadata, putting on captions, annotations, and attributions). The teachers in turn train the participating pupils. In most schools, participating pupils were 10–14 years old, and the number of pupils was such that the teachers could manage the group when reviewing foodway stories and also while on outdoor activities.

Each pupil in consultation with the teacher chooses a topic to research. The topic could be about a plant product, e.g. a fruit. The pupil then goes and gathers all available information in consultation with parents and relatives. The pupil takes photos and videos of various activities related to the food. The pupil then writes an essay about his or her topic, making illustrations where possible, and brings it to the teacher for assessment. The length of this essay is flexible, and it is written in a language agreed upon with the teacher. The pupil gets feedback from the teacher to improve the essay. The pupil also works with the teacher to write captions for selected photos.

Table 10.1 Basic information of the study communities in Kenya and Tanzania where the foodways documentation protocol was applied

Country	Kenya	Kenya	Kenya	Kenya	Tanzania
County/district	Kilifi, Kwale	Kakamega	Baringo	Narok	Ngorongoro
Sub-county/division	Kilifi South, Kaloleni, Rabai (Kilifi), Msambweni, Kinango (Kwale)	Shinyalu	Tiaty	Loita	Loliondo
Study sites	Mijikenda (9 sites): *Kaya* Kauma, *Kaya* Chonyi, *Kaya* Jibana, *Kaya* Kambe, *Kaya* Ribe, *Kaya* Rabai, *Kaya* Fungo (in Kilifi), *Kaya* Mtswakara, *Kaya* Kinondo (in Kwale)	2 sites: Shihuli, Muraka	2 sites: Chemolingot, Churo	2 sites: Entasekera, Olmesutye	2 sites: Waso, Loliondo
Dominant ethnicity(ies), if appropriate	Mijikenda (9 communities): Kauma, Chonyi, Jibana, Kambe, Ribe, Rabai, Giriama, Duruma, Digo	Isukha	Pokot (Eastern)	Loita Maasai	Loita Maasai
Size of the case study/project area (hectare)[1]	25,000	10,000	116,084 (Churo/Amaya and Ribkwo wards)	20,000	20,000
Number of direct beneficiaries (persons)[2]	50,000	40,000	35,799	3000	3000
Geographic coordinates (longitude and latitude)	• 3°37′14.0″S 39°44′10.0″E (Kauma) • 3°48′15.6″S 39°40′54.4″E (Chonyi) • 3°50′15.0″S 39°40′10.0″E (Jibana) • 3°51′49.0″S 39°39′07.0″E (Kambe) • 3°53′49.0″S 39°37′58.0″E (Ribe) • 3°55′55.0″S 39°35′46.0″E (Rabai) • 3°47′55.0″S 39°30′52.0″E (Fungo) • 3°59′54.0″S 39°31′25.0″E (Mtswakara) • 4°23′36.0″S 39°32′41.0″E (Kinondo)	• 0°12′27.7″N 34°47′38.5″E • 0°14′50.7″N 34°45′10.0″E	• 0°58′51.5″N 35°58′17.9″E • 0°46′25.2″N 36°24′29.4″E	• 1°54′22.8″S 35°43′55.5″E • 1°54′22.7″S 35°43′56.9″E	• 2°02′23.2″S 35°42′42.5″E • 2°03′42.4″S 35°33′37.6″E

Note: The superscript numbers indicate that the number of indirect beneficiaries does not apply in this case as it was mainly a documentation exercise. Only those who interacted with the researchers directly can be considered beneficiaries

Table 10.2 Basic information of the study communities in Burkina Faso and Ethiopia where the foodways documentation protocol was applied

Country	Ethiopia	Burkina Faso
State/province	Benishangul-Gumuz	Ioba
District/department	Kamashi	Dissin, Zambo, Dano
Villages/sites/commune	2 main villages: Jalo, Kobi	3 main villages: Zambo, Nakar, and Complan
Size of geographical area (hectare)[1]	90,000	147,000
Dominant ethnicity(ies), if appropriate	Gumuz	Dagara
Size of the case study/project area (hectare)[1]	9000	49,000
Number of direct beneficiaries (persons)[2]	2000	50,000
Geographic coordinates (longitude and latitude)	9°30′16.8"N 35°51′51.5"E (Kamashi zone, Aba Ciotte)	10°57′10.8"N 2°59′10.4"W (Banfaré area)

Note: The superscript numbers indicate that the number of indirect beneficiaries does not apply in this case as it was mainly a documentation exercise

The teachers read all essays, looking at not only the content, but also the grammar and style. The teachers are also the custodians of cameras. They keep all records, including copies of completed pupils' essays. Teachers also ensure that the pupil's name, age, class, school, date, and topic appear on the top of the pupil's essay and that all pages are numbered.

The following are some key topics pupils covered during their interactions with family members:

1. Local names of the food
2. Description of the food
3. Description of where it is found and/or grown
4. Management in the field
5. The process of harvesting or gathering
6. Preparation or recipes
7. Cultural aspects: myths, beliefs, taboos, etc.
8. Other uses and any other information
9. Seasonal availability
10. Trends in utilisation
11. Tools and equipment

The pupils are also guided by questions about what, who, why, when, where, and how, at every stage of compiling their essays.

The multidisciplinary team also plays a key role by providing scientific input including nutritional information and scientific names. The team also ensures that the same information is archived and shared with a national repository according to terms agreed upon with the community. The team works closely with the community

Table 10.3 The diversity of edible species and habitats associated with each community where the foodways documentation was applied (source: prepared by authors, 2021)

Community	Main sources of livelihood	Target environment/landscape	Domesticated animals	Food animals in natural environment	Cultivated food plants only	Wild and semi-cultivated food plants	Total number of species
Mijikenda (Kenya)	Crop cultivation, gathering wild food	Sacred forests (*Kaya*) with farm-lands around them	6 (1.6%)	170 (44%)	35 (9.1%)	181 (46.9%)	386 (100%)
Gumuz (Ethiopia)	Crop cultivation, gathering, hunting	Medium forest, bamboo forest, and bushland with villages and farmlands	6 (3%)	53 (26.4%)	53 (26.4%)	95 (47.3%)	201 (100%)
Dagara (Burkina Faso)	Crop cultivation, gathering, hunting	Medium forest and bushland, with villages and farmlands with natural fruit trees	8 (3.8%)	34 (16.2%)	59 (28.1%)	117 (55.7%)	210 (100%)
Isukha (Kenya)	Crop cultivation	Community farmlands with rich agricultural biodiversity	8 (5.9%)	9 (6.7%)	72 (53.3%)	54 (40%)	135 (100%)
Pokot (Kenya)	Pastoralism	Semi-arid land with high wild diversity	7 (3.5%)	35 (17.6%)	22 (11.1%)	142 (71.4%)	199 (100%)
Loita Maasai (Kenya and Tanzania)	Livestock keep-ing, minimal crop production	Community-managed forest (Loita) and bushland with rich plant and animal diversity	4 (2.6%)	4 (2.6%)	42 (27.6%)	106 (69.7%)	152 (100%)

Note: "Animals" refers to both land and water animals used for food

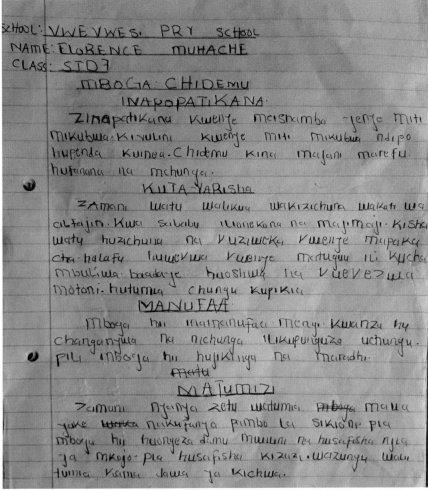

Fig. 10.6 A pupil's essay on Chidemu (Vernonia sp., a leafy vegetable) written in Swahili (photo credit: P. Maundu, 2017)

to identify opportunities for using their knowledge and food resources to improve their livelihoods. Standard consent procedures are followed when dealing with the community.

A typical pupil's essay is provided in Fig. 10.6. In all cases, and as expected, pupils gave priority to their main foods. In the pastoralist Pokot for example, the community gave priority to livestock diversity and food gathered in the wild, while those in the agricultural Isukha gave priority to cultivated foods. Among the foods of the Loita Maasai, milk (the most important Maasai food) had the highest frequency constituting over half of the 450 essays recorded by the two schools involved. About 1000 essays were written by the pupils from the Mijikenda schools over 18 months.

The main challenges faced concerned language and the use of cameras. Maasai pupils preferred to write in their local language (Maasai), while Mijikenda pupils preferred Swahili. This meant engaging translators. In some schools, photos were not taken as school teachers would not entrust pupils with cameras. In most communities, documentation by schools was augmented by simultaneous documentation work by trained community youth. The combined information from the community and pupils resulted in a rich document about food resources from the local landscape.

Some potential outputs of the foodways documentation can be:

- An inventory of traditional foods of a community and indigenous knowledge about them
- Identification of season of availability
- Identification of local recipes
- Identification of food taboos, beliefs, and ceremonial use
- Recognition of the value of a food species to a community

It is evident from Table 10.1 that the people relied on many species obtained from the immediate natural environment. Dependence on wild resources seemed to be particularly high among pastoralists and communities with gathering practices.

The foodways documentation among the Isukha and the Pokot was published in three volumes which are now accessible on the internet (Maundu et al. 2013c,b,c,d).

Follow-Up Development and Conservation Activities

Documentation should not be a stand-alone activity. It should pave the way for community development activities based on the local knowledge and resources. Such activities help communities to treasure their knowledge and resources. The following three cases demonstrate the kind of development projects that can follow documentation.

3   Cases Studies

Each of the three cases presented here has documentation, development, and conservation aspects. In all cases, foodways documentation has proved useful in the development and execution of the activities.

3.1   Case Study 1: African Leafy Vegetables (ALVs) Research and Promotion

Phase I

The main drive to bring back African leafy vegetables (ALVs) to the Kenyan menu started in 1995, and involved a consortium of institutions brought together by the International Plant Genetic Resources Institute (IPGRI). Key partners were the National Museums of Kenya (NMK) and local universities. During the first phase, 1995–1999, local production systems, associated indigenous knowledge, and local use of the ALVs were documented using many aspects of the foodways documentation method. The documentation in this phase not only laid a base for future work but also selected 24 priority vegetables (out of 220 documented) for further research and promotion (Fig. 10.7).

At the end of the first phase, key areas of follow-up were identified, including increasing awareness in the general public; collecting seeds of priority species; improving seed systems; developing protocols for cultivation; linking farmers to markets; documenting recipes; and conducting nutritional analysis. This guided the preparation of the second phase.

Phase II

In this phase between 2001 and 2006, the multi-stakeholder group of partners was expanded to include government ministries, national research institutions, non-governmental organisations (NGOs), community-based organisations (CBOs), and more universities. The team embarked on improving people's perceptions of local vegetables through promotional activities based on scientific evidence. Nairobi City, the commercial hub, was strategically chosen for promotional activities, firstly

Fig. 10.7 The orange-fruited African nightshade (*Solanum villosum*), one of the priority leafy vegetables (photo credit: P. Maundu, 2012)

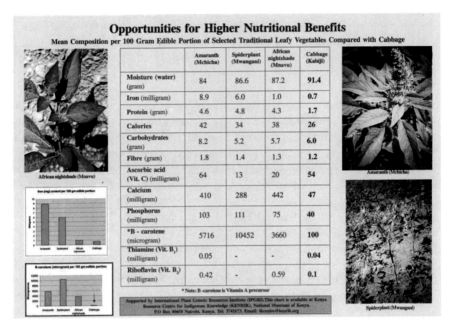

Fig. 10.8 A leaflet used in the promotion of traditional vegetables in Kenya (source: P. Maundu, 2002)

because it has people from various ethnic communities (hence has a ready market for a variety of species) and secondly, what goes on in the capital eventually trickles down to the rural areas. Farmers in Nairobi's peri-urban agricultural areas were particularly targeted for the production of traditional vegetables. The fact that most traditional vegetables are highly nutritious proved helpful during promotional campaigns. Promotional activities took the form of cooking demonstrations, street and media campaigns, field days, posters, and leaflets providing recipes and nutritional benefits of the food (Figs. 10.8 and 10.9). By 2003, a change was noticeable. The negative attitude among vegetable consumers had begun to crumble. Traditional vegetables started to flood through both formal and informal markets, opening income-generating opportunities for farmers and traders who are mainly women. Farmers had a reason to conserve the species and their landscapes (Maundu et al., 2008).

Alongside promotional work, the project gathered germplasm, some of which underwent selection work for improvement at the World Vegetable Centre facilities in Arusha, Tanzania. Improved seeds were distributed to farmers and seed companies. Farmers around Nairobi were given agronomic support and linked to formal markets, initially Uchumi Supermarkets, one of the main supermarket chains at the time.

By 2006, consumption had increased, creating unprecedented demand (Gotor & Irungu, 2010). Most households in and around the city of Nairobi incorporated traditional vegetables in their diets. Supermarkets started to sell the vegetables

Fig. 10.9 Promotional campaigns on the streets of Nairobi in June 2004 (photo courtesy: National Museums of Kenya, 2004)

while street vendors flooded some streets with the vegetables (Figs. 10.10 and 10.11). Restaurants started to include vegetables in their menus. As attitudes had changed and stigma turned into pride, the once neglected traditional vegetables became an area of interest for development agencies and training and research institutions. Production and marketing slowly moved from city suburbs into rural areas. Up to 17 traditional vegetable species were now regularly sold in local markets. The years that followed saw many organisations, including NGOs and universities, come to the scene and implement projects on ALVs.

Fig. 10.10 Traditional vegetables flood the streets of Nairobi after the 2003–2006 campaign (photo credit: P. Maundu, 2006)

Post-Phase II Years and the Impact of the ALV Promotion

From 2008 to 2011, Bioversity International in partnership with the National Museums of Kenya and other organisations used a similar research and promotion approach in Kitui County located about 100 km east of Nairobi. One area within the county (Mwingi South) was chosen as the control site and Kitui Central as the intervention site. A baseline situation was established before the intervention. Neither the "control" nor the "intervention" markets sold traditional vegetables like the African nightshade and spider plant. After a year of promotional work in Kitui Central, a marked difference was noticeable. Attitudes in Kitui Central had changed, and more traditional vegetables were being cultivated and sold there. The diversity of vegetables in diets had also increased. The African nightshade, amaranth, and spider plant became mainstream traditional vegetables in Kitui Central

Fig. 10.11 Traditional vegetables had made a come-back by 2003 with most supermarkets stocking them. Tuskys Supermarket, Ongata Rongai (photo credit: P. Maundu, 2006)

(Kariuki et al., 2013) and only appeared on the "control" side at least a year later and also after an end-line intervention.

The above work is a clear demonstration that, with awareness and promotion, neglected and underutilised species can again become part of local diets, contributing to nutrition, health, and incomes, particularly among peri-urban women. The involvement of a multidisciplinary research and development team is a key factor to success.

This work on ALVs has been evaluated by a number of individuals and organisations interested in specific aspects of the outcomes. An impact assessment conducted in 2006 and 2007 positively attributed increased production, marketing, and consumption to the work of the consortium. The women-dominated industry saw increased incomes for peri-urban women producers and city vendors. Nearly two-thirds (62.5%) of households growing ALVs increased their income from the sale of ALVs, while nearly half increased their consumption of ALVs. The assessment further found that interest in ALVs had increased among researchers, producers, traders, development workers, and consumers. Seeds had become more accessible, and increased cultivation of previously wild species was also observed. Traditional knowledge about the vegetables and associated practices had been revived (Gotor & Irungu, 2010).

3.2   Case Study 2: Bottle Gourd (Kitete) Conservation

The bottle gourd project, which was implemented between 2000 and 2002, used the foodways documentation method to identify the diversity of bottle gourds, local management practices, and related knowledge and cultural practices in Kitui County and surrounding counties (Morimoto & Maundu, 2002). Using the documented information, the project raised awareness among local communities to create economic, cultural, and livelihood opportunities by increasing bottle gourd cultivation and incentivising maintenance of the unique bottle gourd diversity.

A community group, Kyanika Adult Women Group in Kitui County, was concerned about losing the diversity and indigenous knowledge (IK) of an important cultural species due to the introduction of plastic containers, and solicited support from scientists at Bioversity International and the National Museums of Kenya. Group members started to collect seeds of different landraces and to document the indigenous knowledge of all the diverse bottle gourd shapes and types—initially from group members, then from other communities in Kitui County, and finally from more distant counties where they had contacts. Scientists guided the process. The group was trained on documentation protocols including use of the camera and tape recorder. One of the greatest innovations of the group was initiating an activity whereby group members formed pairs and then travelled to their original parental homes, often in faraway and diverse places, to interview elderly parents and relatives and collect bottle gourd seeds from them. The purpose of the pairing was for the women to be seen as belonging to a group and also to provide support to one another. Upon returning, members shared their experiences with group members and compiled the information in a 700-page compendium about the species. Nearly 200 bottle gourd types (landraces) referred to by approximately 70 different local names were collected and catalogued for propagation in community fields (Morimoto & Maundu, 2002; Morimoto et al., 2005, 2008; Morimoto, 2010).

The group members grew both edible gourds and those used as containers, and shared seeds with neighbours and friends. They also established a *kitete* community museum within the village. The museum serves as a centre for storing and distributing bottle gourd seeds, and also for learning about the bottle gourd. Visitors include school children, scientists, and bottle gourd enthusiasts (Morimoto et al., 2010). In addition, the group, in partnership with the different stakeholders, learned skills for decorating the *kitete* and *nzele* (a half gourd used as a bowl) and began selling these items to restaurants and tourist centres (Fig. 10.12).

Fig. 10.12 Members of Kyanika Women's Group displaying decorated bottle gourds while youth from a visiting school look on (photo credit: P. Maundu, 2012)

3.3 Case Study 3: Introducing Modern Technology to Educate Consumers on Diversifying Diet with Local Foods—The ADD-IT System

The ALV approach provided lessons that were later applied to promote dietary diversity in general. A study conducted by the team in Kitui County examined the link between food diversity (in the farm and wild landscapes), dietary diversity (at the household), and nutrition and health status of individuals. A set of 10 villages (the intervention villages) were exposed to promotional activities such as community seed fairs and seed exchanges, traditional food cooking demonstrations, and awareness-raising on nutrition for a period of 2 years (2009–2011). In another set of 10 villages (control villages) separated from the first by a mountain ridge, no promotional activities were carried out until after the two-year period had ended. The study found that diversity (agro-biodiversity) at individual households in the intervention villages increased significantly between the baseline and final survey, while agro-biodiversity did not change much in the control villages. Dietary diversity at the household level showed the same trend. The study could not, however, establish a connection between the increased diversity at household level and any impacts on people's health (Maundu, 2011; Herforth et al., 2019). The changes observed were attributed to the intervention activities. This work found that, despite communities having good access to diverse foods, poor dietary habits due to lack of awareness are sometimes to blame for poor nutrition. As was the case with ALV case study, awareness and education changed individuals' food choices, especially incorporating more local foods in diet.

Based on the above experience, a new tool (an app) called the Agrobiodiversity Diet Diagnosis Interventions Toolkit (ADD-IT) was developed by the Alliance of Bioversity International and the International Center for Tropical Agriculture (CIAT) in partnership with the National Museums of Kenya and the Tokyo University of Agriculture, Japan.

The app utilises a newly developed, validated food frequency questionnaire (FFQ) to understand the food consumption patterns of the subject. Using a number of inbuilt datasets including a database on locally available foods, nutrient composition, and recommended daily nutrient allowance, the app is able to determine over- and underconsumption of specific food groups. It then provides nutritional education using an inbuilt nutrition education facility which guides the subject through the steps to take in order to stay within the country's dietary guidelines, but with specific emphasis on local foods (Induli, 2018; Irie et al., 2019).

ADD-IT thus (1) helps the subjects visualise their dietary patterns (i.e. deficiencies and excesses), (2) provides relevant real-time nutrition education based on what foods are available locally (in the community) at the time (season), and (3) stores the data in a web-based database system that can be used for further assessment.

The application is built mainly with community health workers in mind (Fig. 10.13) but will help any respondent(s) including households to understand their consumption habits. The overall data collected by the app at community level may help planners to initiate more targeted programmes to counter malnutrition or specific deficiencies at the community level. The application can also be used to accumulate new information on local food diversity.

The creation of this tool involved public-private sector cooperation. A private company provided the technical support to create the app, and scientists provided the databases on which it runs, while health workers and extension staff tested the app and provided feedback for further improvement.

4   Discussion

This chapter has highlighted the significance of the foodways documentation approach in the identification of useful local food resources in local landscapes. We have also seen that foodways documentation itself is not the end, but rather a stepping stone to developing programmes that can benefit the community. Benefits in the form of income or better diets can act as incentives for communities to conserve and sustainably manage the local landscapes where these foods grow. In the case of the African leafy vegetables, the vegetable habitats and the species

Fig. 10.13 A community health volunteer (CHV) in Vihiga County (holding tablet) uses a prototype ADD-IT system tool to assess dietary patterns of the respondent (looking on) and provide feedback (photo credit: Y. Morimoto, 2019)

themselves were more appreciated when an economic aspect was introduced, and this gave the vegetables and their habitats some form of protection. The increased demand for the vegetables as a result of promotional activities spurred production and trade, thus creating more opportunities for income (especially for women who dominate vegetable cultivation and trading activities) and household nutrition.

As seen earlier, the foodways documentation methodology helps us understand food in relation to other aspects of community culture. It is therefore less useful in situations where a community has significantly lost its culture and diversity. Community involvement, especially of the youth, is a key feature of the method. Much can be achieved when school children are mobilised as they can handle different aspects of the foodways. With the advancement and spread of digital tools and forums such as Facebook, Twitter, and YouTube, it will become even easier for the youth to share information on foodways.

The educational aspect of the ADD-IT tool builds upon data gathered from the foodways documentation. Aside from conservation and development opportunities, foodways documentation also opens up opportunities for nutrition research, such as the 24-h dietary recall and the food frequency questionnaire. The methodology is still being improved as the authors continue to test it in different production systems and cultures. Improvement is needed in the area of managing the massive amount of data that is churned out. Partnerships between communities and the private sector can create good synergies.

This chapter has also shown how the methodology has been applied at different levels of diversity—including the landscape, ecosystem, species, and variety levels.

At the ecosystem level, foodways documentation has led to the empowerment of pupils and communities situated around the biodiversity-rich sacred *Kaya* forests of the Mijikenda and the Loita Maasai Forest, leading them to appreciate the diversity of their immediate ecosystem. In these cases, the communities have chosen to conserve wild areas as forestland rather than converting it to agricultural land—a case of balancing conservation and development objectives. Such well-conserved ecosystems (Maundu et al., 2001; NMK, 2008) are perfect examples of the multiple dimensions of health as they are not only a source of food and herbs for nutrition and health, but also a source of mental and spiritual health (many cultural ceremonies are held in the forest). The forests offer livelihood security through benefits such as water sources, non-timber forest products (e.g. honey), climate change mitigation and adaptation, and general community livelihoods (Maundu et al., 2001). As is with most conservation work, there are trade-offs. Increased selection of a species is likely to narrow the genetic pool of the species. However, with good management plans (e.g. promoting both wild and improved (selected) varieties on the basis of culinary strengths or preferences and cultural uses), these shortcomings can be overcome.

Moreover, this chapter has shown that community participation and leadership in documentation, as well as awareness on the value of local foods and cultural knowledge and practices, help in achieving ecosystem health, human health, and sustainability within socio-economic production landscapes (SEPLs) management (Bergamini et al., 2014; Morimoto et al., 2015; Mijatovic et al., 2018). Therefore, the effectiveness of SEPL management with respect to foodways can be measured through the following:

- Food diversity at the landscape level (diversity of production habitats, food diversity, household dietary diversity)
- Community benefits (income, health including mental and physiological, nutrition, cultural identity, general welfare)
- Community leadership and level of participation and interest (local champions, community support, and involvement in decision-making)
- Youth involvement (mentorship, youth in leadership)

4.1 COVID-19 Effects

In Kenya, the COVID-19 pandemic has had an unprecedented impact on the well-being of the people, not so much from infections and mortalities, but economically, socially, and psychologically. Lockdowns, restriction on movement, loss of jobs, higher cost of living, etc. have all taken a heavy toll on the people's livelihoods. However, rural people who earn most of their livelihood from their immediate natural environment have been impacted much less by the pandemic than those in urban areas. This is a further demonstration that SEPLs can serve as the basis of resilience in local communities in the event of catastrophes and especially those affecting human health (Morimoto et al., 2015; Bedmar Villanueva et al., 2018; Dunbar et al., 2020). While this enhanced protection may be a result of plants providing a physical barrier and low population density, it could also be due to the healthy and diverse foods and herbal medicines in the environment, which the local people ingest daily, as well as their psychological status (Matsuda & Morimoto, 2020). The Maasai of Loita, for example, use diverse plants in soups and milk (key foods of the community) for health (Maundu et al., 2001). Some of the plants used for respiratory diseases in infusions are known to have antiviral and anti-inflammatory properties. The authors have observed that the COVID-19 pandemic has made many Kenyans more conscious about their health, and therefore it has pushed the demand for traditional vegetables, fruits, spices, and herbal teas higher, forcing many to grow their own healthy foods (Xinhua, 2020). These cases, and particularly the ALV and the ADD-IT projects, have a direct relevance to the COVID-19 pandemic as nutrition security is a short-term goal.

5 Conclusion

The case studies described herein have demonstrated not only the importance of foodways documentation in the sustainable management of rural landscapes, but also its role in community development, including ensuring human and ecosystem health. Foodways documentation is fundamental in development programmes that aim to maximise the benefits of landscape resources. Such programmes require multi-stakeholder action in collaboration with local communities as demonstrated in the three case studies. The studies have shown that multiple benefits to the communities serve as incentives for conservation and sustainable use (Maundu & Morimoto, 2011). Biodiverse community-managed landscapes such as the Loita Maasai Forest and *Kaya* sacred forests of the Mijikenda give multiple benefits including resilience for food security, nutritional and health benefits, good mental health, and prevention of infectious diseases like COVID-19. The cases provide insights on how foodways documentation approaches can contribute to sustainable management of natural resources, achievement of global biodiversity and sustainable development goals, and good health for all.

References

Bedmar Villanueva, A., Morimoto, Y., Maundu, P., Jha, Y., Otieno, G., Nankya, R., Ogwal, R., Leles, B., & Halewood, M. (2018). Perceptions of resilience, collective action and natural resources management in socio-ecological production landscapes in East Africa. In UNU-IAS & IGES (ed.) *Sustainable use of biodiversity in socio-ecological production landscapes and seascapes (SEPLS) and its contribution to effective area-based conservation* (, pp.14–25), Satoyama initiative thematic review, vol. 4. United Nations University Institute for the Advanced Study of Sustainability

Bergamini, N., Dunbar, W., Eyzaguirre, P., Ichikawa, K., Matsumoto, I., Mijatovic, D., Morimoto, Y., Remple, N., Salvemini, D., Suzuki, W., & Vernooy, R. (2014). *Toolkit for the indicators of resilience in socio-ecological production landscapes and seascapes*. UNU-IAS, Biodiversity International, IGES & UNDP.

Dunbar, W., Subramanian, S. M., Matsumoto, I., Natori, Y., Dublin, D., Bergamini, N., Mijatovic, D., González Álvarez, A., Yiu, E., Ichikawa, K., Morimoto, Y., Halewood, M., Maundu, P., Salvemini, D., Tschenscher, T., & Mock, G. (2020). Lessons learned from application of the "Indicators of resilience in socio-ecological production landscapes and seascapes (SEPLS)" under the Satoyama initiative. In O. Saito, S. M. Subramanian, S. Hashimoto, & K. Takeuchi (Eds.), *Managing socio-ecological production landscapes and seascapes for sustainable communities in Asia* (pp. 93–116). Science for Sustainable Societies, Springer.

Google. (2015). Google maps, SW Burkina Faso, viewed 5 December 2015. Retrieved from https://www.google.co.ke/maps/@11.0135609,-2.9839067,10.99z/data=!5m1!1e1.

Google. (2016). Google Maps, Kenya, viewed 31 October 2016. Retrieved from https://www.google.co.ke/maps/@-0.275945,38.5953367,7z.

Google. (2020). Google maps, West, Central and East Africa, viewed 1 October 2020. Retrieved from https://www.google.co.ke/maps/@6.7311926,16.9259445,4.98z.

Gotor, E., & Irungu, C. (2010). The impact of Bioversity International's African leafy vegetables programme in Kenya. *Impact Assessment and Project Appraisal, 28*, 41–55.

Hama-Ba, F., Parkouda, C., Thiombiano, N., Maudi, P., & Diawara, B. (2016). *Recettes a base de legumes du couvert forestier livret de recettes traditionnelles a base de legumes du couvert forestier en pays Dagara au Burkina Faso*. CIFOR, Bioversity International and IRSAT. Retrieved from https://www.cifor.org/publications/pdf_files/others/6586-booklet.pdf.

Hama-Ba, F., Sibiria, N., Powell, B., Ickowitz, A., Maundu, P., & Diawara, B. (2017). Micronutrient content of wild vegetable species harvested in forested and non-forested areas in Southwest Burkina Faso. *Forest Research, 6*(3).

Herforth, A., Johns, T., Creed-Kanashiro, H. M., Jones, A. D., Khoury, C. K., Lang, T., Maundu, P., Powell, B., & Reyes-García, V. (2019). Agrobiodiversity and feeding the world: More of the same will result in more of the same. In K. S. Zimmerer & S. de Haan (Eds.), *Agrobiodiversity: Integrating knowledge for a sustainable future* (Strüngmann forum reports book 24) (pp. 185–211). The MIT Press.

Induli, I. (2018). *Better data for better nutrition*. Bioversity International, viewed 11 February 2021. Retrieved from https://www.bioversityinternational.org/news/detail/better-data-for-better-nutrition/.

Induli, I. (2019a). *Integrated participatory approach to improving dietary diversity*. Panorama Solutions, viewed 11 February 2021. Retrieved from https://panorama.solutions/en/solution/integrated-participatory-approach-improving-dietary-diversity.

Induli, I. (2019b). *Enhancing use and conservation of food plant resources around sacred Kaya forests of the Mijikenda people of Kilifi, Mombasa and Kwale counties, Kenya*. Panorama Solutions, viewed 11 February 2021. Retrieved from https://panorama.solutions/en/solution/enhancing-use-and-conservation-food-plant-resources-around-sacred-kaya-forests-mijikenda.

Induli, I., Morimoto, Y., & Maundu, P. (2020). *Researchers and entrepreneurs bring back forgotten gems: Underutilized crops transformed into healthy snacks*. Biodiversity International, viewed 11 February 2021. Retrieved from https://www.bioversityinternational.org/news/detail/researchers-and-entrepreneurs-bring-back-forgotten-gems-underutilized-crops-transformed/.

Irie, K., Morimoto, Y., & Maundu, P. (2019). Development of a new integrated ICT system tool (Agrobiodiversity and diet diagnosis for interventions toolkit: ADD-IT) for better decision-making in nutrition interventions. *Journal of the Agricultural Society of Japan, 1651*, 27–35. (in Japanese).

JAICAF. (2020). *Introducing a new business using popping cereal technology in Kenya*. YouTube, viewed 11 February 2021. Retrieved from https://www.youtube.com/watch?v=8oM_D2bxaYQ.

Johns, T., Powell, B., Maundu, P., & Eyzaguirre, P. B. (2013). Agricultural biodiversity as a link between traditional food systems and contemporary development, social integrity and ecological health. *Journal of the Science of Food and Agriculture, 93*(14), 3433–3442.

Kariuki, L., Maundu, P., & Morimoto, Y. (2013). Some intervention strategies for promoting underutilized species: Case of local vegetables in Kitui District, Kenya. *Acta Horticulturae, 979*, 241–248.

Kenya National Bureau of Statistics (KNBS). (2014). *Kenya demographic and health survey 2014*, Nairobi, Kenya, viewed 10 November 2021. Retrieved from https://dhsprogram.com/pubs/pdf/fr308/fr308.pdf.

Kilifi Udamaduni Conservation Group. (2010). In P. Maundu, Y. Morimoto, & E. Towett (Eds.) *Mboga za watu wa Pwani (Vegetables of coastal people of Kenya)*. Bioversity International, viewed 10 November 2021. Retrieved from http://www.bioversityinternational.org/fileadmin/_migrated/uploads/tx_news/Mboga_za_Watu_wa_Pwani_1515.pdf (in Kiswahili).

Matsuda, H., & Morimoto, Y. (2020). Linkage between natural resource, agriculture, nutrition, and health toward sustainable society. *AgriBio* (pp. 67–70), Hokurikukan, New Science Co. Ltd. (in Japanese).

Maundu, P. M., Ngugi, W. G., & Kabuye, H. S. C. (1999a). *Traditional food plants of Kenya*. National Museums of Kenya.

Maundu, P. M., Njiro, E. I., Chweya, J. A., Imungi, J. K., & Seme, E. N. (1999b). The biodiversity of traditional leafy vegetables in Kenya. In J. A. Chweya & P. B. Eyzaguirre (Eds.), *The biodiversity of traditional leafy vegetables* (pp. 51–83). International Plant Genetic Resources Institute.

Maundu P., Berger, D., Saitabau, C., Nasieku, J., Kipelian, M., Mathenge, S., Morimoto, Y., & Hoft, R. (2001). *Ethnobotany of the Loita Maasai: Towards community management of the forest of the lost child—Experiences from the Loita ethnobotany project*. People and Plants Working Paper 8. UNESCO.

Maundu, P. (2002). *Opportunities for higher nutritional benefits*. Leaflet, International Plant Genetic Resources Institute.

Maundu, P., Grum, M., Morimoto, M., Eyzaguirre, P., & Johns, T. (2008). Diet diversification through local foods: Experiences from traditional vegetable promotion work in Kenya. In Poster presented at the international forum EcoHealth 2008, Merida, Yucatan, Mexico, 1–5 December.

Maundu, P., Achigan-Dako, E., & Morimoto, Y. (2009). Biodiversity of African vegetables. In C. M. Shackleton, M. W. Pasquini, & A. W. Drescher (Eds.), *African indigenous vegetables in urban agriculture* (pp. 65–104). Earthscan.

Maundu, P. (2011). Managing agricultural biodiversity for better nutrition and health, improved livelihoods and more sustainable production systems in sub-Saharan Africa: Case studies from Benin, Kenya, and South Africa. International Development Research Centre, viewed 10 November 2021. Retrieved from http://hdl.handle.net/10625/47124.

Maundu, P., & Morimoto, Y. (2011). Reconciling genetic resources and local knowledge conservation and livelihoods enhancement in research and development: Experiences of biodiversity International in sub-Sharan Africa. *Biodiversity, 9*(1), 56–60.

Maundu, P., Muiruri, P., Adeka, R., Ombonya, J., Morimoto, Y., Bosibori, E. Kibet, S., & Odubo, A. (2013c). *Safeguarding intangible cultural heritage: A photobook of the traditional foodways of the Isukha and East Pokot communities of Kenya*. UNESCO, viewed 10 November 2021. Retrieved from https://www.bioversityinternational.org/e-library/publications/detail/safeguarding-intangible-cultural-heritage-a-photobook-of-traditional-foodways-of-the-isukha-and-eas/.

Maundu, P., Kapeta, B., Muiruri, P., Morimoto, Y., Bosibori, E., Kibet, S., & Odubo, A. (2013a). *Safeguarding intangible cultural heritage: Traditional foodways of the East Pokot community of Kenya*. UNESCO, viewed 10 November 2021. Retrieved from https://www.bioversityinternational.org/e-library/publications/detail/safeguarding-intangible-cultural-heritage-traditional-foodways-of-the-east-pokot-community-of-kenya/.

Maundu, P., Muiruri, P., Adeka, R., Ombonya, J., Morimoto, Y., Bosibori, E. Kibet, S., & Odubo, A. (2013d). *Safeguarding intangible cultural heritage: Traditional foodways of the Isukha community of Kenya*. UNESCO, viewed 10 November 2021. Retrieved from https://www. bioversityinternational.org/e-library/publications/detail/safeguarding-intangible-cultural-heri tage-traditional-foodways-of-the-isukha-community-of-kenya/.

Maundu, P., Bosibori, E., Kibet, S., Morimoto, Y., Odubo, A., Kapeta, B., Muiruri, P., Adeka, R., & Ombonya, J. (2013b). *Safeguarding intangible cultural heritage: A practical guide to documenting traditional foodways*. UNESCO, viewed 10 November 2021. Retrieved from https://www.bioversityinternational.org/e-library/publications/detail/safeguarding-intangible-cultural-heritage-a-practical-guide-to-documenting-traditional-foodways/.

Mijatovic, D., Hodgkin, T., Pawera, L., Meldrum, G., Jarvis, D., Sthapit, S., Zira, S., Morimoto, Y., Maundu, P., Álvarez, A., Tarraza, A., Palikhey, E., Azhdari, G., Gruberg, H., Wakkumbure, L., Salimi, M., Estrada-Carmona, N., Shabong, R., & Maneerattanachaiyong, S. (2018). *Assessing agrobiodiversity a compendium of methods*. Platform for Agrobiodiversity Research.

Mkuu, R. S., Epnere, K., & Chowdhury, M. (2018). Prevalence and predictors of overweight and obesity among Kenyan women. *Preventing Chronic Disease, 15*(4), E44.

Morimoto, Y., & Maundu, P. (2002). Community-based documentation of indigenous knowledge (IK), awareness and conservation of cultural and genetic diversity of the Bottle Gourd (*Lagenaria siceraria*) in Kitui District, Kenya. In Poster presented at the Deutscher Tropentag workshop 2002, Challenges to organic farming and sustainable land use in the tropics and subtropics, University of Kassel, Witzenhausen, Germany, 9–11 October 2002. Retrieved from https://www.tropentag.de/2002/abstracts/posters/158.pdf.

Morimoto, Y., Maundu, P., Fujimaki, H., & Morishima, H. (2005). Diversity of landraces of the white-flowered gourd (*Lagenaria siceraria*) and its wild relatives in Kenya: Fruit and seed morphology. *Genetic Resources and Crop Evolution, 52*(6), 737–747.

Morimoto, Y., Maundu, P., & Eyzaguirre, P. (2008). The bottle gourd: Its origin, travels round the globe with man and cultural influences on its fruit characteristics. In Poster presented at the International Symposium of Jack Harlan II, UC Davis, California, USA, 13–19 September 2008, viewed 10 November 2021. Retrieved from https://cgspace.cgiar.org/bitstream/handle/10 568/67528/Morimoto_thebottlegourd.pdf?sequence=1&isAllowed=y.

Morimoto, Y., Maundu, P., Tumbo, D., & Eyzaguirre, E. (2010). How farmers in Kitui use wild and agricultural ecosystems to meet their nutritional needs (Kenya). In C. Belair, K. Ichikawa, B. Y. L. Wong, & K. J. Mulongoy (Eds.) *Sustainable use of biological diversity in socio-ecological production landscapes: background to the 'Satoyama Initiative for the benefit of biodiversity and human well-being*. Secretariat of the Convention on Biological Diversity, CBD Technical Series no. 52, pp. 67–72.

Morimoto, Y. (2010). Countering local knowledge loss and landrace extinction in Kenya: The case of the bottle gourd (*Lagenaria siceraria*), Case 37. In L. Maffi, & E. Woodley (Eds.) *Biocultural diversity conservation: A global source book* (pp. 33–35). Earthscan Publications Ltd.

Morimoto, Y., Maundu, P., Mijatovic, D., Bergamini, N., & Eyzaguirre, P. (2015). Assessing farmers' perception of resilience of socio-ecological production landscapes in central and eastern Kenya. In UNU-IAS & IGES (Ed.), *Enhancing knowledge for better management of socio-ecological production landscapes and seascapes (SEPLS)* (Satoyama initiative thematic review) (Vol. 1, pp. 96–106). United Nations University Institute for the Advanced Study of Sustainability.

NMK. (2008). *The sacred Mijikenda Kaya forests*. National Museums of Kenya, Nairobi, viewed 10 November 2021. Retrieved from https://whc.unesco.org/uploads/nominations/1231rev.pdf.

Xinhua. (2020). Kenyan herbs exports to increase in 2020 amid COVID-19 pandemic. *Xinhua*, 14 November, viewed 14 May 2021. Retrieved from http://www.xinhuanet.com/english/2020-11/14/c_139514478.htm.

The opinions expressed in this chapter are those of the author(s) and do not necessarily reflect the views of UNU-IAS, its Board of Directors, or the countries they represent.

Open Access This chapter is licenced under the terms of the Creative Commons Attribution 3.0 IGO Licence (http://creativecommons.org/licenses/by/3.0/igo/), which permits use, sharing, adaptation, distribution and reproduction in any medium or format, as long as you give appropriate credit to UNU-IAS, provide a link to the Creative Commons licence and indicate if changes were made.

The use of the UNU-IAS name and logo, shall be subject to a separate written licence agreement between UNU-IAS and the user and is not authorised as part of this CC BY 3.0 IGO licence. Note that the link provided above includes additional terms and conditions of the licence.

The images or other third party material in this chapter are included in the chapter's Creative Commons licence, unless indicated otherwise in a credit line to the material. If material is not included in the chapter's Creative Commons licence and your intended use is not permitted by statutory regulation or exceeds the permitted use, you will need to obtain permission directly from the copyright holder.

Chapter 11
Multi-stakeholder Approach to Conserving Agricultural Biodiversity and Enhancing Food Security and Community Health During the COVID-19 Pandemic in Kampong Cham, Cambodia

Jeeranuch Sakkhamduang, Mari Arimitsu, and Machito Mihara

Abstract Agricultural biodiversity plays a vital role in enhancing food security and human health. Sustainable agriculture practices that conserve soil and water can result in good environmental and human health. In view of this, a project on capacity-building for sustainable agricultural practices targeting extension officers was implemented between September 2017 and February 2021 in Kampong Cham Province, Cambodia, by the Institute of Environmental Rehabilitation and Conservation (ERECON), Japan. The project involved government agencies, educational institutes, NGOs, and farmers, and employed a multi-stakeholder approach to promote sustainable farming practices among local farmers and enable conditions for the sale of agricultural products with low chemical inputs, especially in a province where agrochemical application is prevalent. A questionnaire survey, key informant interviews, focus group discussions, and observations from farmers were used for programme monitoring. Farmers reported that soil quality was improved after applying compost, and more beneficial insects were found after integrated pest management techniques were applied. The amount of agrochemicals applied to farmlands decreased compared to usage before the project start, implying that the project was successful in promoting sustainable agriculture in the province. During the COVID-19 pandemic, communities in the project areas are struggling to cope

J. Sakkhamduang (✉)
Southeast Asia Office, Pathum Thani, Thailand

Institute of Environmental Rehabilitation and Conservation, Tokyo, Japan

M. Arimitsu
Institute of Environmental Rehabilitation and Conservation, Tokyo, Japan

M. Mihara
Institute of Environmental Rehabilitation and Conservation, Tokyo, Japan

Tokyo University of Agriculture, Japan, Faculty of Regional Environment Science, Tokyo, Japan

© The Author(s) 2022

M. Nishi et al. (eds.), *Biodiversity-Health-Sustainability Nexus in Socio-Ecological Production Landscapes and Seascapes (SEPLS)*, Satoyama Initiative Thematic Review, https://doi.org/10.1007/978-981-16-9893-4_11

with food and health insecurity. The intervention has helped communities become more resilient during this hard time. After 3 years, many of the approximate 1500 farmers involved in the project are applying organic fertilisers and enhancing agricultural biodiversity in their farmlands. This case is a grassroots-level activity, but the concept of multi-stakeholder activities for agricultural biodiversity conservation can be replicated in other areas of Cambodia for achieving the sustainable development goals.

Keywords Sustainable agricultural practices · Agricultural biodiversity · Food security · Multi-stakeholder approach · COVID-19

1 Introduction

In Cambodia, the application of chemical fertilisers and pesticides has significantly increased to promote agricultural productivity. However, the inappropriate use of agrochemicals, such as overuse and application without sufficient knowledge, has caused various problems to human and environmental health (Paavo & Sergiy, 2015). Although agricultural productivity has increased temporarily, environmental issues including soil degradation, water contamination from agrochemicals, and water quality degradation such as eutrophication have resulted. Several studies have proposed that one way to avoid the adverse impacts of agrochemicals on humans and the environment is sustainable agriculture (Lotter, 2003; Crowder et al., 2010; Eyhorn, 2007).

Agriculture plays an important role in Cambodia by ensuring food security at community and national levels as well as in providing income opportunities. In order to improve agricultural production, Cambodia imports chemical fertilisers from other countries. Recently the chemical fertiliser and pesticide markets have been growing rapidly, and application has become quite common among Cambodia's farmers (Ministry of Environment (MoE), 2004). In 2001, chemical fertilisers and pesticides were imported into Cambodia in the amount of 45,335 tons and 200 tons, respectively (Ministry of Environment (MoE), 2004), while in 2010, 245,854 tons of chemical fertiliser and 1357 tons of pesticides were imported into the country (Ministry of Environment (MoE), 2010; FAOSTAT, 2021). Many Cambodian farmers believe that increased agricultural production can only be achieved by using modern inputs, especially agrochemical products. This practice contributes to increased farm products, but also increases farm expenditure and the risks for human and environmental health (Smith et al., 1990).

Kampong Cham Province is located in the central region, or plain zone according to the topographical classification, of Cambodia with a population of 895,763 people within 215,923 households (Royal Government of Cambodia, 2019) (Table 11.1). Among the five provinces in the plain zone, Kampong Cham has the highest poverty severity index and poverty gap index of 3.34 and 9.28, respectively (JICA, 2010). Almost all of the area in the province is non-forest land (agricultural and residential areas), as shown in Fig. 11.1. Agricultural lands can be categorised into two distinct

Table 11.1 Basic information of the study area

Country	Cambodia
Province	Kampong Cham
District	All 9 districts in the province
Municipality	1
Size of geographical area (hectare)	459,400
Number of direct beneficiaries (persons)	1530
Number of indirect beneficiaries (persons)	899,791
Dominant ethnicity(ies), if appropriate	Khmer
Size of the case study/project area (hectare)	459,400
Geographic coordinates (latitude, longitude)	11° 59′ 0″ N, 105° 27′ 0″ E

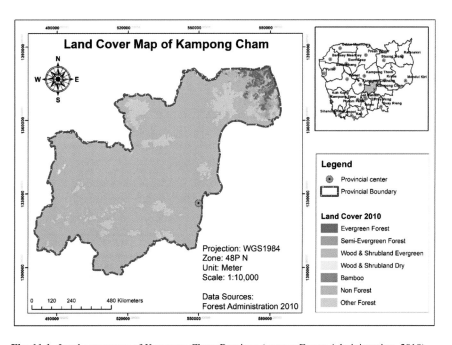

Fig. 11.1 Land cover map of Kampong Cham Province (source: Forest Administration, 2010)

topographical regions: lowlands and uplands. Lowland areas mainly support rice farming interspersed with field crops, vegetable gardens, and fruit trees. Upland areas are mainly used for rubber (*Hevea brasiliensis*) plantations, maize (*Zea mays*), cassava (*Manihot esculenta*), soybeans (*Glycine max*), mung beans (*Vigna radiata*), peanuts (*Arachis hypogaea*), sesame (*Sesamum indicum*), sugar cane (*Saccharum officinarum*), and fruit trees (Ministry of Agriculture, Forestry and Fisheries (MAFF), 2012). Similar to large-scale farmers in the country, farmers in Kampong Cham rely heavily on agrochemical products to increase farm production. EuroCham (2016) reported that 123,871 and 79,328 households in the province applied inorganic fertilisers and chemical pesticides, respectively, to their farmlands.

Although many farmers realise the adverse impacts of agrochemicals to human and environmental systems, they lack knowledge related to sustainable or alternative farming practices. Meanwhile, the agricultural extension officers who are responsible for enhancing farmers' knowledge are simply not enough in number compared to their assigned areas (de Silva et al., 2014). Cambodia has suffered prolonged internal conflict that has ruined the educational system and social economy across the country, such that capacity-building of extension officers has not been implemented. Hence, interventions from change agents such as NGOs, civil society, or educational institutions are essential to tackle this problem.

To address the aforementioned issues, the Institute of Environmental Rehabilitation and Conservation (ERECON) initiated the Project on Promoting Sustainable Agricultural Conditions for Poverty Reduction in Kampong Cham Province (September 2017 to February 2021) to enhance the capacity of extension officers and farmers, with a multi-stakeholder approach as the key approach. The project aimed to promote sustainable agricultural conditions through various forms of education for change agents (agricultural extension officers) and farmers in the province. The project covered all ten districts of the province. The specific objectives included:

1. To build capacity on sustainable agriculture based on the cyclic use of natural resources in the Provincial Department of Agriculture, Forestry and Fisheries (PDAFF) and ten District Department of Agriculture, Forestry and Fisheries (DDAFF) of the province, including dissemination skills and knowledge ranging from soft (knowledge, skills) to hard (facilities) measures
2. To promote sustainable farming practices based on the cyclic use of natural resources by local farmers
3. To promote conditions for the sale of agricultural products with low chemical inputs

Although human health was not a main objective of the programme from the beginning, from April 2020 to February 2021 amidst the COVID-19 global pandemic, programme stakeholders observed that due to achievement of objective number two, some changes had occurred related not only to environmental but also human health.

2 Methodology

2.1 Multi-stakeholder Approach

Hemmati (2002) described the multi-stakeholder process as a tool to promote better decision-making by ensuring that the views of the main actors concerned about a particular decision are heard and integrated at all stages of dialogue and consensus building. The process takes the view that everyone involved in the process has a valid view and relevant knowledge and experience to bring to the decision-making.

Table 11.2 Stakeholders and their roles in the project

Stakeholder group	Number of participants from each stakeholder group	Role
Provincial Department of Agriculture, Forest and Fisheries (PDAFF) officers	5	Attend trainings and workshops and transfer knowledge to model and DG farmers
District Department of Agriculture, Forest and Fisheries (DDAFF) officers	25	Attend trainings and workshops and transfer knowledge to model and DG farmers
Model Farmers and District Group (DG) Members	500	Attend workshops and transfer knowledge to general farmers
General Farmers	1000	Attend workshops and transfer knowledge to other farmers and neighbours
Royal University of Agriculture (RUA)	3	Prepare technical content for trainings and workshops and giving trainings and lectures in workshops
Kampong Cham National Institute of Agriculture (KNIA)	2	Prepare technical content for trainings and workshops, giving trainings and lectures in workshops, and programme monitoring
Institute of Environmental Rehabilitation and Conservation (ERECON)	10	Prepare technical content for trainings and workshops, giving trainings and lectures in workshops, programme monitoring, and programme facilitating

The approach aims to create trust between actors and solutions that provide mutual benefits (win-win). The approach is people-centred, with everyone involved taking responsibility for the outcome. When inclusive and participatory approaches are used, stakeholders have a greater sense of ownership of decisions made, and are thus more likely to comply with them. Stakeholders of the project included the Royal University of Agriculture (RUA), Cambodia; Provincial Department of Agriculture, Forestry and Fisheries (PDAFF), District Department of Agriculture, Forestry and Fisheries (DDAFF), Kampong Cham; Kampong Cham National Institute of Agriculture (KNIA); and farmers' groups for promoting sustainable agriculture. The number of persons and the roles of each stakeholder group are presented in Table 11.2.

Trainings and workshops covered topics ranging from sustainable agricultural practices to product handling and marketing channels and were provided to extension officers and farmers by ERECON staff and RUA and KNIA lecturers, as presented in Table 11.3. Contents of training and workshops were prepared and delivered to PDAFF and DDAFF officers by ERECON staff and RUA and KNIA lecturers. The trained officers gave feedback on the content and delivery methods to lecturers and ERECON staff. The feedback was considered and incorporated into outreach materials before it was used with farmers, as shown in Fig. 11.2. The

Table 11.3 Topics of trainings/workshops and number of participants

Duration	Topic of training/workshop	Participants
January 2018–June 2018	Improvement of soil fertility Soil conservation practices Composting, pellet compost Liquid fertiliser Multicropping Agroforestry	PDAFF and DDAFF officers (30) Model and DG farmers (500) General farmers (1000)
14–17 May 2018	Technical training in Thailand for agricultural extension officers	PDAFF and DDAFF officers (20) KNIA staff (1) ERECON staff (1)
January 2019–June 2019	Integrated pest management Biopesticide making	PDAFF and DDAFF (30) Model and DG farmers (500) General farmers (1000)
13–18 May 2019	Technical training in Thailand for agricultural extension officers	PDAFF and DDAFF officers (20) KNIA staff (1) ERECON staff (3)
January 2020–June 2020	Irrigation techniques Drip irrigation Mini sprinkle planning and installing	PDAFF and DDAFF officers (30) Model and DG farmers (500) General farmers (1000)
September 2020–November 2020	Post-harvesting, product handling, and marketing channels	PDAFF and DDAFF officers (30) Model and DG farmers (500) General farmers (1000)

roughly 1500 farmers who attended the trainings and workshops were expected to share the knowledge gained with their neighbours. Agricultural materials, such as compost boxes and tanks for making liquid fertiliser and biopesticide, were distributed to all farmers along with guidebooks on sustainable agriculture practices. Networks for safe agricultural products were formed, and marketing channels for the products were introduced to farmers in the final year of the project through workshops and trainings. Small shops for selling low agrochemical input products were set up in the ten districts of the province, with products sold at the stalls coming from farmers who participated in the project. The main products included leafy vegetables, tomatoes, eggplants, cucumbers, etc.

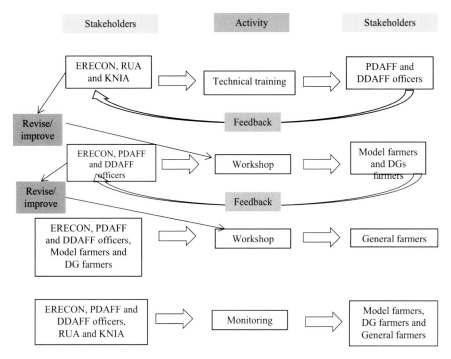

Fig. 11.2 Stakeholder involvement in the Project on Promoting Sustainable Agricultural Conditions for Poverty Reduction in Kampong Cham Province (from October 2017 to February 2021) (source: authors)

2.2 Data Collection and Analysis

A questionnaire survey, semi-structured interviews, focus group discussions, and programme monitoring through farm data records and farm visits by DDAFF officers and KNIA and ERECON staff were the methods used to collect data. Moreover, the concept of citizen science (McKinley et al., 2017; Ryan et al., 2018) was applied by asking farmers to observe and make note of changes observed in their farmlands after converting to organic fertilisers. DDAFF officers were assigned to make monthly visits to farmers participating in the project, make records of natural fertilisers produced and utilised in farmlands, and write monthly reports. To capture any changes following project implementation, 100 DG farmers who participated in the project were randomly selected out of a total of 500 DG farmers to conduct baseline and end-line surveys. Descriptive statistics were used to analyse the data collected.

3   Results

From the first year of the project until project termination, more than 50 workshops were conducted for agricultural officers and farmers in the project area as shown in Table 11.4. Trainings and workshops for PDAFF and DDAFF officers were conducted at the provincial office. Officers working in the districts had to travel to the provincial capital to attend. For DG and general farmers, workshops were held at model farmers' houses or at village meeting places as shown in Figs. 11.3 and 11.4. The average distance travelled to attend workshops was around 4 km.

According to the baseline and end-line surveys, 78.0% of respondents were male with an average age of 49.62 years old. Regarding educational attainment, 44.0% graduated primary school, and 39.0 percent graduated secondary school. The average number of family members was 5.12 persons, and the average years of residency

Table 11.4 Number of workshops and participants during project implementation

Participants	No. of participants per training or workshop	Number of workshops per topic per district (10 districts)	Duration of each workshop (in days)	Number of topics	Total number of trainings/ workshops organised
PDAFF and DDAFF officers (30)	32	1	2	4	4
Model farmers and DGs (500)	25	2	1	4	20
General farmers (1000)	30–50	2–3	1	4	32

Fig. 11.3 Workshop on composting

Fig. 11.4 Training on biopesticides (photo courtesy of the Institute of Environmental Rehabilitation and Conservation (ERECON), 2018a, b)

Table 11.5 General information on respondents (source: survey data)

Item		Respondents (of total 100)	
		(Value)	(%)
Gender (persons)	Male	78	78.0
	Female	22	22.0
Average age (years)		49.62	
Educational background of respondent (persons)	No schooling	4	4.0
	Primary	44	44.0
	Secondary	39	39.0
	High school	12	12.0
	College/university	1	1.0
Average number of family members (persons)		5.12	
Average years of residence	(years)	43.04	
Average total cultivation area (ha)		1.19	
Irrigation of farmland (households)	Fully	72	72.0
	Partially	26	26.0
	Not at all	2	2.0
Average annual income from agricultural activity (in KHR)		8,385,414.14	
Average annual income from non-agricultural activity (in KHR)		2,247,838.38	
Average annual total income (in KHR)		10,633,252.53	

in the respective districts were 43.04 years. The average cultivated land size was 1.19 ha with 72% fully irrigated. Main crops grown by the respondents included rice, cassava, corn, fruit trees, and vegetables.

Even without any significant change in cropping area, end-line surveys showed that the farmers' average annual income from agricultural activity increased by 49.9% or 3,683,141.41 KHR (Cambodian Riel, around 905 USD) from the year that the project started, pointing to an increased efficiency in the existing land use (Table 11.5). Income from non-agricultural activities also grew by 45.2% or

1,015,079.76 KHR (249.5 US$) over the 2-year period (2018–2020). One of the reasons for this increase was family members gaining income working in garment factories or working abroad sending remittances back to their families.

3.1 Changes in Agricultural Practices

Farmers who reported using chemical pesticides decreased overall by 9.0% from 90.0 to 81.0 percent. Of these, farmers using herbicides decreased by 9.0% from 54.0 to 45.0%, and those using insecticides decreased by 11.0% from 78.0 to 67.0%. Meanwhile, farmers who reported using organic pesticides increased by 39.0% from 23.0 to 62.0%. Similarly, farmers using chemical fertilisers decreased by 5.0 percentage points from 87.0 to 82.0% (Table 11.6).

The farmers who did not apply agrochemicals prior to joining the project stated reasons including lands being relatively small (less than one hectare) and a lack of any significant insect pest problems. They considered application of agrochemicals to be unnecessary.

Likewise, comparing with the baseline survey, end-line survey results showed that the number of farmers using organic fertilisers increased by 11.0% from 85.0 to 96.0%. Among organic fertilisers, the biggest increase was seen for compost, rising by 74.0% from 7.0 to 81.0%, followed by liquid bio-fertiliser rising 25.0% from 4.0 to 29.0%, and green manure with a 28.0% increase from 6.0 to 34.0%. Rice husk and/or bran went up 20.0% from 13.0 to 33.0%, rice straw 12.0% from 13.0 to 25.0%, and other organic fertilisers 6.0% from 2.0 to 8.0%. Of all organic fertilisers, only animal manure saw a relative decrease in use, falling by 7.0% from 80.0 to 73.0% due to farmers using it to make compost instead of applying it directly on their fields (Table 11.7).

The reason behind this drastic increase in organic fertiliser use was the farmers' lack of knowledge in making organic fertilisers, such as compost or liquid fertiliser, before the project started. Prior to the project, the most prevalent practice was applying cow manure directly to farmlands, which carried the risk of pathogenic bacteria such as *Escherichia coli* leaking into water sources. Through participating in the project, farmers gained knowledge on how to make and apply compost and liquid

Table 11.6 Changes in application of fertilisers and pesticides (source: survey data)

Item	Respondent (100)	
	Baseline (%)	End line (%)
Using chemical fertilisers	87.0	82.0
Using chemical pesticides	90.0	81.0
Herbicides	54.0	45.0
Fungicides	32.0	32.0
Insecticides	78.0	67.0
Other	1.0	3.0
Using organic pesticides	23.0	62.0

Table 11.7 Changes in organic fertiliser application (source: survey data)

	Respondents (100)	
Item	Baseline (%)	End line (%)
Using organic fertiliser	85.0	96.0
Rice straw	13.0	25.0
Rice husk and/or bran	13.0	33.0
Green manure	6.0	34.0
Animal manure	80.0	73.0
Compost	7.0	81.0
Pellet compost	1.0	6.0
Liquid bio-fertiliser	4.0	29.0
Other	2.0	8.0

fertiliser to their farmlands. During the second year of the project, when some farmers started to grow low input or organic vegetables, compost became a necessary input for them. Many farmers changed their attitudes towards organic fertilisers, especially compost, after observing the good practices of their neighbours.

However, some farmers still apply chemical fertilisers mixed with compost in an attempt to get a quicker and higher yield than with compost alone. Some farmers have continued to decrease the amount of chemical fertilisers used while increasing the amount of compost. Chemical pesticides are still widely used in the project areas due to organic pesticides showing a lower performance than that of chemical pesticides, especially for farmers who grow large amounts of vegetables. Changing the perception of farmers concerning the adoption of organic fertiliser application by modelling the change proved to be very effective. In this case, the model farmers played important roles in sharing knowledge and using their farms as places for demonstration and showcasing of sustainable agricultural practices to their neighbours and visitors.

Regarding changes in soil conditions and diversity in farms, semi-structured interviews were conducted with ten model farmers. Excerpts from the interviews are presented below:

First thing I did is reduce chemical fertiliser and pesticide; after participate in the training, I clearly realise that applying chemicals is impacting health and the environment and costs much more money. That is the reason why I changed to using compost for which materials can be found around the house, and it's cheaper. Second is I changed the traditional habit of growing; I used to grow crops for only household consumption and without much taking care. But now, that habit is changed, and I look after my crops and soil well. I can see the soil condition is improved after applying compost. —Farmer A

I can say, it (crop diversity) is increased 30% in my farms. I cultivate many types of vegetables, and I can produce foods for the year round. —Farmer B

Crop diversity in my farm is really increased. More than 50 dragon fruit trees are increased, and I plan to add another 30 trees this year. Moreover, I grow more crops for household consumption, like eggplant, tomato, lemongrass, and other herbs, etc. I also raise chickens, ducks and some catfish to gain more income. —Farmer A

Since I started applying bio-pesticide to my farmland, I found out there are many beneficial insect populations that have significantly increased. I noted that there are bees, spiders, golden bugs, and frogs, etc. —Farmer B

I noticed that after I started applying bio-pesticide, some beneficial insects appeared more, like bees and spiders. There are also birds and some types of reptiles coming to eat worms, ants, and other pests. Those insects and small animals never showed up when I used chemical pesticide. —Farmer C

Most farmers interviewed reported noticing improvements in soil conditions, such as soil being more porous, softer, and darker in colour compared to before applying organic fertilisers. Farmers also reported that they increased crop diversity in their farmlands with multiple benefits, such as producing for household consumption, for the market, or to use as safety nets (in the case of animal husbandry). Moreover, more beneficial insects were found in farmlands after farmers stopped applying chemical pesticides and began using biopesticides made from plants and herbs in their farmlands.

3.2   Impact of COVID-19

Kampong Cham is one of the provinces with reported cases of COVID-19, with 118 cases as of 4 May 2021 (Ministry of Health (MoH), 2021). Due to the global pandemic, several planned workshops and trainings had to be postponed because the Royal Government of Cambodia restricted all events involving more than 50 participants. The postponed activities have caused a delay in project implementation, especially in the flood-prone areas along the Mekong River, which are normally difficult to access during the rainy season.

During this hardship, several organisations working in the country decided to hold workshops, seminars, or meetings using internet platforms to connect people. However, most of the project beneficiaries are farmers living in rural and remote areas with limited access to infrastructure such as tap water or electricity, let alone smartphones to connect to the internet. Most of the farmers participating in the project own mobile phones, but only 25% own smartphones. Some farmers reported that although at least one of their family members owns a smartphone, they themselves do not know how to use it.

Face-to-face communication and social gatherings are still the most effective means of communication with farmers, especially in remote areas. Therefore, local stakeholders, especially agricultural extension officers, model farmers, and district group members, play important roles in project monitoring and keeping the majority of farmers engaged in the project via farm visits and mobile phone conversations.

A questionnaire survey for DG and model farmers and semi-structured interviews were conducted to obtain updates on farmers' situations during the pandemic, especially on their feelings about their own health, food security, and mental health. A total of 58 DG farmers completed the survey between January and May 2021. Results are shown in Table 11.8.

Table 11.8 Farmers' feelings towards health, food security, diversity in their farms, and mental health during COVID-19 (source: survey data)

Items	Respondents (58)	
After applying organic fertilisers to your farmland, do you think you are healthier than before?	Healthier54	No difference4
Did you increase types of plants/livestock in your farmland more than before?	More diverse49	No difference9
Do you think you have more food security than before?	More food security53	No difference3
Do you think you have chances to meet with other farmers from different villages/communes more often than before?	Yes57	No difference1
COVID-19 concerns	Number of respondents (multiple responses allowed)	
How does COVID-19 affect your everyday life?		
I worry that I or my family members are going to contract COVID-19.	55	
I cannot attend meetings or workshops conducted by government agencies or NGOs as usual.	33	
I do not want to go to public places such as pagodas, markets, or meeting halls as I might contract COVID-19.	50	
I feel less confident about consuming foods from markets as it might be contaminated with COVID-19.	41	
My agricultural products cannot be sold as before due to COVID-19.	33	
I feel depressed about the situation of COVID-19 and economic situation.	38	
How do you cope with such concerns or fears?		
I and my family members always wear facial masks and wash our hands often, especially when we go out.	56	
I attend meetings or workshops as little as possible.	35	
I avoid going to public places.	46	
I eat more of my own foods, such as vegetables or livestock that I grow rather than buying raw materials from markets.	40	
I try to contact other farmers or agricultural officers to find other market channels.	30	
I talk and share my feelings with neighbours and farmers I know from the project.	35	
Do you think sharing your concerns or fears with neighbours and other farmers from the project can reduce your stress caused by the COVID-19 situation?		
Yes, very much	32	
A little	23	
Not at all	3	

Some responses from interviews with farmers are presented below:

Since COVID-19, I am only confident in consuming food collected from my own farm. And I do not usually buy from outside, only meats and other necessary ingredients. I can rely on my farm to survive during the pandemic. —Farmer D

Since the COVID-19 outbreak, I am more confident eating my own food from my farm rather than going out to eat or buying vegetables from unknown sources. Because I only apply compost, liquid fertiliser and bio-pesticide I have learnt from ERECON about my crops. Our community members also know how to prevent themselves from contracting COVID-19 by following the Ministry of Health's announcements. Many of them diversified their farms to meet household consumption and meet the market demand. But at the same time, their products do not sell well, because of the reduction of buyers/consumers. —Farmer E

Most of the farmers interviewed stated that during the pandemic, they felt more confident consuming their own products, believing that they could depend on their own crop diversity. Meanwhile, they felt less stress after talking and exchanging information among community members. Results of interviews clearly show that sustainable agricultural practices lead to healthy soil and healthy and safe foods and contribute to human health, both of the individual and the community.

4 Challenges and Opportunities

4.1 Challenges

During the 3 years of the project intervention, the number of farmers using organic fertilisers has increased, but the use of chemical fertilisers mixed with compost is still prevalent, although the amount of chemical fertiliser is small compared to compost. Project staff and agricultural extension officers suggested farmers to gradually decrease the amount of chemical fertilisers and increase the amount of compost. Moreover, the biopesticides introduced by the project were not highly adopted by some farmers, especially those engaged in mono-crop farming, as they are less effective compared to chemicals. Positive human and institutional impacts were seen as extension officers gain knowledge, skills, and experience in agricultural extension and advisory services from agricultural development projects/programmes (Ke & Babu, 2018). However, the number of agricultural extension officers assigned to work in each district is still small considering the size of the areas for which they are responsible. This challenge is also discussed in the work of Sothath and Sophal (2010), who note that the average of extension service support in Cambodia at the district level is 5000 households per extension worker. This limitation is an obstacle for farmers to obtain adequate knowledge of sustainable farming practices.

4.2   Opportunities

Sustainable organic farming has become one of the province's agricultural extension policies. The project has contributed to provincial and district agricultural strategic plans by promoting the production of low agrochemical input or organic products by farmers in the province for supply to local markets. Moreover, the project has also enhanced food security at the household and community level through increased crop diversity and productivity. Organic farming and safer foods are well accepted by relevant stakeholders, especially customers in the province. Several farmers involved in the project produced low agrochemical input products or organic products, which earned 20 to 25 percent higher a price than that of conventional products. Figure 11.5 shows the organic vegetable farm belonging to one of the DG farmers, which received a net house from another project as part of an integrated pest management (IPM) technique promotion activity. The farmer has successfully applied IPM techniques and demonstrated the benefit of selling low-chemical-input products at higher prices. This case serves as a good example that can motivate more farmers to follow a similar path. Enhanced engagement of youth and educational institutes was also observed. The project stakeholders included two educational institutes, one located in the capital city of Phnom Penh and the other located in Kampong Cham Province, the Kampong Cham National Institute of Agriculture, which actively engaged in the project by conducting farm visits and collaborating with farmers using their farms as experimental sites for students. Sharing knowledge with university students made the farmers more confident in their farming practices. Moreover, the project also encouraged young farmers to get involved and use their knowledge of and access to social media to support older farmers in keeping up with news on techniques and other information related to sustainable agricultural practices.

During project implementation, we observed the active involvement of women. Several of them supported the extension officers during trainings and workshops by

Fig. 11.5 Organic vegetable farm of a DG member

Fig. 11.6 A farmer showing her organic garden to visitor (photo courtesy of the Institute of Environmental Rehabilitation and Conservation (ERECON), 2021 (Fig. 11.5) and Institute of Environmental Rehabilitation and Conservation (ERECON), 2019 (Fig. 11.6))

sharing their knowledge and first-hand experiences in applying organic fertilisers or other changes in their farms with other farmers as well with visitors from universities and government agencies (Fig. 11.6). Several PDAFF and DDAFF officers acknowledged that through this project, they acquired new knowledge and techniques on sustainable farming practices used both in Cambodia and abroad. The technical trainings in Thailand, in particular, motivated many officers to develop conditions for sustainable farming practices in their own responsible areas. Moreover, the knowledge dissemination design that required officers to give lectures to farmers in the workshops gave junior officers the opportunity to hone their teaching and communication skills instead of only observing senior officers perform.

5   Conclusion

After 3 years of project implementation, several changes and developments in terms of agricultural practices and the capacities of officers and farmers were observed. According to monitoring reports and farm visits, farmers produced and applied compost, liquid fertilisers, and biopesticides in their farmland instead of agrochemical products. The changes resulted in an increase in agricultural biodiversity in farmlands and enhanced food security and human health.

The project has contributed to provincial and district agricultural strategic plans by promoting the production and sale in markets of low-agrochemical-input or organic products by farmers in the province. Although the progress of marketing organic products was hindered by COVID-19, knowledge in marketing and networks were formed. Moreover, consumers in the Kampong Cham capital city became aware of organic farming networks through the promotional event held by PDAFF.

Through its multi-stakeholder approach, the project addressed Target 2.4 for Goal 2 of the SDGs, by promoting sustainable food production systems to increase productivity and production, and improve land and soil quality. The project covered the whole province encompassing various landscapes ranging from uplands to lowlands and river ecosystems. The sustainable agricultural practices promoted and applied by farmers in these landscapes have helped to conserve the soil and water ecosystems and have resulted in improved human health.

The project also contributed to Target 12.2 for Goal 12, by promoting sustainable management and efficient use of natural resources through composting and producing organic fertilisers. Generally, after rice is harvested from paddies, rice straw is collected and used as animal fodder. However, farmers always burnt the remaining rice stumps as a convenient way to clear the land before the next growing season. Burning residue not only creates smoke and particulate matter (PM2.5), but also kills microorganisms and animals in the soil (Ajay et al., 2019). Using farm residues to make compost to apply to fields is a good example of cyclic use of resources for sustainable production and consumption. Although more labour by farmers is required to make and apply compost than needed for the use of chemical fertilisers, the practices result in a better environment and improved human health both at the individual and community levels.

Farmers who converted their agricultural practices from applying agrochemical products to using organic fertilisers observed several changes in their farmlands. For instance, some noted the improvement of soil conditions, such as more porous soils that are darker in colour and contain more soil organisms. The integrated pest management also allowed more beneficial insects to live, helping control insect pests. Crop diversity in farmlands enhanced the food security of farmers, which has proven to be vital especially during the COVID-19 pandemic.

It is anticipated that understanding and awareness of sustainability, especially with regard to consumption and production, will spread to the public in Kampong Cham Province through the farmers' network. We also expect that the outcomes from this project will be scaled up and applied to broader areas of Cambodia.

References

Ajay, K., Kushwaha, K. K., Singh, S., Shivay, Y. S., Meena, M. C., & Nain, L. (2019). Effect of paddy straw burning on soil microbial dynamics in sandy loam soil of indo-gangetic plains. *Environmental Technology & Innovation, 16*, 100469.

Crowder, D. W., Northfield, T. D., Strand, M. R., & Snyder, W. E. (2010). Organic agriculture promotes evenness and natural pest control. *Nature, 466*, 109–112.

de Silva, S., Johnston, R., & Sellamuttu, S. S. (2014). *Agriculture, irrigation and poverty reduction in Cambodia: Policy narratives and ground realities compared*, CGIAR Research Program on Aquatic Agricultural Systems, Penang, Malaysia, Working Paper AAS-2014-3.

EuroCham. (2016). *Agriculture and agro-processing sector in Cambodia, taking stock: A detailed review of current challenges and investment opportunities in Cambodia*. EuroCham, AgriProject Reporting, v. 46.

Eyhorn, F. (2007). Organic farming for sustainable livelihoods in developing countries: The case of cotton in India. PhD dissertation, Department of Philosophy and Science, University of Bonn.

FAOSTAT. (2021). Pesticides trade, viewed 15 October 2021. Retrieved from https://www.fao.org/faostat/en/#data/RT.

Forest Administration. (2010). *Land cover map of Kampong Cham*. Ministry of Agriculture, Forestry and Fisheries.

Institute of Environmental Rehabilitation and Conservation (ERECON). (2018a). Workshop on composting.

Institute of Environmental Rehabilitation and Conservation (ERECON). (2018b). Training on bio-pesticides.

Institute of Environmental Rehabilitation and Conservation (ERECON). (2021). Organic vegetable farm of a DG member.

Institute of Environmental Rehabilitation and Conservation (ERECON). (2019). A farmer showing her organic garden to visitor.

JICA. (2010). Kingdom of Cambodia, study for poverty profiles in the Asian region, final report, Japan International Cooperation Agency, OPMAC Corporation.

Ke, S. O., & Babu, S. C. (2018). *Agricultural extension in Cambodia: An assessment and options for reform*. IFPRI discussion paper 01706. International Food Policy Research Institute

Hemmati, M. (2002). *Multistakeholder processes for governance and sustainability*. Earthscan.

Lotter, D. W. (2003). Organic agriculture. *Journal of Sustainable Agriculture, 21*(4), 59–128.

McKinley, D. C., Miller-Rushing, A., Ballard, H. L., Bonney, R., Brown, H., Cook-Patton, S. C., Evans, D. M., French, R. A., Parrish, J. K., Phillips, T. B., Ryan, S. F., Shanley, L. A., Shirk, J. L., Stepenuck, K. F., Weltzin, J. F., Wiggins, A., Boyle, O. D., Briggs, R. D., Chapin, S. F., III, . . . Soukup, M. A. (2017). Citizen science can improve conservation science, natural resource management, and environmental protection. *Biological Conservation, 208*, 15–28.

Ministry of Agriculture, Forestry and Fisheries (MAFF). (2012). Annual report of agriculture for 2011–2012.

Ministry of Environment (MoE). (2004). National profile on chemicals management in Cambodia.

Ministry of Environment (MoE). (2010). *Unsustainable agricultural practices* (unpublished handout, Cambodia National Environmental Performance Assessment (EPA) at Sunway Hotel.

Ministry of Health (MoH). (2021). *COVID-19 statistics*, communicable disease control department, viewed 4 May 2021. Retrieved from https://covid19-map.cdcmoh.gov.kh/?fbclid=IwAR2 9DjPwYCLbkqtreXbXnRgsAVjOqEAA0HnX96NQnLLdeT4XKep_PDwH-3U (In Khmer language).

Paavo, E., & Sergiy, Z. (2015). *Cambodian agriculture in transition: Opportunities and risks*. World Bank Group.

Royal Government of Cambodia. (2019). *General population census of the kingdom of Cambodia 2019*. National Institute of Statistics, Ministry of Planning.

Ryan, S. F., Adamson, N. L., Aktipis, A., Andersen, L. K., Austin, R., Barnes, L., Beasley, M. R., Bedell, K. D., Briggs, S., Chapman, B., Cooper, C. B., Corn, J. O., Creamer, N. G., Delborne, J. A., Domenico, P., Driscoll, E., Goodwin, J., Hjarding, A., Hulbert, J. M., . . . Dunn, R. R. (2018). The role of citizen science in addressing grand challenges in food and agriculture research. *Proceedings of the Royal Society B, 285*(1891), 20181977. https://doi.org/10.1098/rspb.2018.1977

Smith, S. J., Schepers, J. S., & Porter, L. K. (1990). Assessing and managing agricultural nitrogen losses to the environment. *Advances in Soil Science, 14*, 1–32.

Sothath, N., & Sophal, C. (2010). Agriculture sector financing and services for smallholder farmers, NGO Forum on Cambodia.

The opinions expressed in this chapter are those of the author(s) and do not necessarily reflect the views of UNU-IAS, its Board of Directors, or the countries they represent.

Open Access This chapter is licenced under the terms of the Creative Commons Attribution 3.0 IGO Licence (http://creativecommons.org/licenses/by/3.0/igo/), which permits use, sharing, adaptation, distribution and reproduction in any medium or format, as long as you give appropriate credit to UNU-IAS, provide a link to the Creative Commons licence and indicate if changes were made.

The use of the UNU-IAS name and logo, shall be subject to a separate written licence agreement between UNU-IAS and the user and is not authorised as part of this CC BY 3.0 IGO licence. Note that the link provided above includes additional terms and conditions of the licence.

The images or other third party material in this chapter are included in the chapter's Creative Commons licence, unless indicated otherwise in a credit line to the material. If material is not included in the chapter's Creative Commons licence and your intended use is not permitted by statutory regulation or exceeds the permitted use, you will need to obtain permission directly from the copyright holder.

Chapter 12
Reducing Commodity-Driven Biodiversity Loss: The Case of Pesticide Use and Impacts on Socio-Ecological Production Landscapes (SEPLs) in Ghana

Yaw Osei-Owusu, Raymond Owusu-Achiaw, Paa Kofi Osei-Owusu, and Julia Atayi

Abstract Ghana's Western North and Central Regions are biodiversity-rich landscapes. Cocoa is a major commodity produced in these two regions, accounting for over 50% of Ghana's cocoa output. As part of the efforts to further improve productivity and ecological health of the landscape, the Government of Ghana initiated the Cocoa Disease and Pests Control Programme primarily to control cocoa pests and diseases, including the use of pesticides. In recent times, however, there has been an upsurge in the use of highly hazardous pesticides (HHPs) that have far-reaching consequences on human and ecological health of the cocoa production landscape. To gain a better understanding of pesticide-use patterns on cocoa farms and address HHP-driven biodiversity loss, Conservation Alliance International (CA) conducted a study within the landscape. The study was based on both qualitative and quantitative research approaches to understand pesticide use and resulting impacts on human and ecological health. In all, 306 cocoa farmers were surveyed. Analysis of the data revealed that about 81% of the cocoa farmers use pesticides to address pests and diseases, causing visible impacts on humans and the environment, including skin irritation, eye irritation, and death of pollinators. Pesticide use was exacerbated by the adverse economic impacts of the COVID-19 pandemic. Policymakers are therefore advised to take steps to phase out HHPs, promote integrated pest management, and tackle the spread of COVID-19 infections.

Keywords Pesticides · Cocoa · Communities · Environment · Health · Landscape · Biodiversity

Y. Osei-Owusu (✉) · R. Owusu-Achiaw · P. K. Osei-Owusu · J. Atayi
Conservation Alliance International, Accra, Ghana
e-mail: yosei-owusu@conservealliance.org; rowusu-achiaw@conservealliance.org;
pkosei-owusu@conservealliance.org; a.julia@conservealliance.org

© The Author(s) 2022

247

M. Nishi et al. (eds.), *Biodiversity-Health-Sustainability Nexus in Socio-Ecological Production Landscapes and Seascapes (SEPLS)*, Satoyama Initiative Thematic Review, https://doi.org/10.1007/978-981-16-9893-4_12

1   Introduction

Cocoa production is a major economic activity in Ghana. Many studies have confirmed that the cocoa sector is the mainstay of the economy, employing about 45.0% of the workforce (Vigneri & Kolavalli, 2018; Gakpo, 2012). It provides 30% of Ghana's total export earnings making it the second largest source of export earnings after mineral exports (World Bank Group, 2018). Since 1990, the Government of Ghana has implemented several policies aimed at reforming the cocoa sector in order to increase the national cocoa output. Notably, in 1999, the government developed a national cocoa development strategy to increase cocoa production from 300,000 metric tons (MT) to 700,000 MT by 2010 (Amoah, 2013). In 2001, the government through its agency, the Ghana Cocoa Board (COCOBOD), designed another programme named the Cocoa Disease and Pests Control (CODAPEC) programme, primarily to assist in the control of cocoa pests and diseases (Essegbey & Ofori-Gyamfi, 2012). The CODAPEC programme provides free spraying of approved pesticides over one acre of cocoa farm per farm to address incidences of diseases and pests in an effort to curb the declining cocoa output (Anang et al., 2013; Boadu, 2014). The programme does not cover the additional acres of farms that are of more than one acre in size. Hence, cocoa farmers are expected to purchase additional approved pesticides to cover the rest of their farms.

Additionally, the programme provided opportunities for enhancing the ecological health of the production landscape by promoting agroforestry systems, integrated pest management (IPM), and conservation of faunal IUCN Red List species (Adjinah & Opoku, 2010). It also made provisions for enhancing the technical capacity of key staff of the Ghana Health Service (GHS), Forest Services Division (FSD), Wildlife Division (WD), and the Environmental Protection Agency (EPA), as well as Protected Area Managers, to monitor the impacts of pesticide application on the health of humans and protected areas. It therefore provided the requisite motivation for conservation practitioners to collaborate with the programme proponents and relevant agencies, including the Ministry of Food and Agriculture (MoFA), to address cases of pesticide poisoning among humans and ecosystems through alternatives, including IPM. IPM is an environmentally sensitive approach to pest and disease management that relies on a combination of common cultural practices (Singh & Prasad, 2016). IPM programmes have a proven track record of significantly reducing the risks of pesticides while improving the health of the ecosystem. Cultural practices such as pruning trees and maintaining a close canopy, among others, are practical ways to prevent pest attacks and also give agronomic benefits to the cocoa trees.

By 2012, the CODAPEC programme had significantly increased national cocoa output from 600,000 MT to a record 1,000,000 MT. This record was 35.7% higher than the full-year record of 740,000 MT, obtained in the 2005/2006 season (ISSER, 2011). Much of the increase in cocoa output was attributed to increased pesticide application on cocoa farms (Afrane & Ntiamoah, 2011). While the increase in output is most welcomed, several studies have suggested that cocoa farmers have become overly dependent on pesticides in addressing disease and pest control, posing enormous risk to humans and the environment within the cocoa production landscape (Conservation Alliance (CA), 2019, 2020; Pesticide Action Network (PAN) UK, 2018). Biodiversity loss within the agricultural portion of the production landscape is mostly driven by pesticide use. Other factors degrading biodiversity in the protected portions of the landscape include illegal logging, agricultural encroachment, and poaching.

Damalas and Eleftherohorinos (2011) indicated that pesticides are widely used in agricultural production to prevent losses due to pests and improve yields. Unfortunately, pesticides reduce the populations of insects, spiders, and birds that naturally control pests and pollinate the crops, through either direct effects or indirect effects (such as lowering the number of flowering weeds visited by insects) (Gill & Garg, 2014). These hazards have been exacerbated by flooding of the markets with highly hazardous pesticides (HHPs) and by abuses—including the use of mixtures of different pesticides, use in higher concentrations than recommended, and application at higher frequencies than required—by most farmers. More worryingly, the applications are carried out without strict adherence to precautionary measures including wearing of appropriate personal protective equipment (PPE). Additionally, current government regulations and control measures in regard to the use of unapproved pesticides including HHPs—such as fines—are not enough of a deterrent and are poorly enforced. This is exacerbated by the lack of a clear policy direction on timelines from the Government of Ghana with respect to phasing out HHPs within the agricultural production landscapes.

The use of pesticides provides a number of benefits, including control of vector-borne diseases (Conservation Alliance (CA), 2020). However, the use of these toxic materials on the production landscape has potential hazards that are of great human and ecological health significance (Conservation Alliance (CA), 2019). Concern has risen in recent years that the current pesticide regulatory system, which is intended to minimise health risk to the general population, may not adequately protect the health of cocoa farmers and the environment (Pesticide Action Network (PAN) UK, 2018). Additionally, the increasing demand for more cocoa for export has resulted in the increased use of pesticides (Conservation Alliance (CA), 2020).

Aimed at cutting down the use of pesticides, Conservation Alliance International (CA), an NGO, carried out a public awareness and education programme on the adverse effects of pesticide abuse within the landscape through radio broadcasts and community public address systems. Awareness on the associated risks of dependency on pesticides could motivate cocoa farmers to press for alternatives. As part of the awareness program, a study was conducted by CA to understand pesticide use and impacts on human health and the environment since pesticide use poses a major challenge in managing the production landscape. Another challenge for managing the production landscape has been weak institutional coordination among stakeholders due to overlapping mandates. There is an opportunity to address this challenge through the local government authority, which is developing a strategic plan to address biodiversity loss and promote sustainable development through a jurisdictional landscape approach that encompasses activities of actors under one central coordination unit to enhance the success of project interventions.

The main goal of this study was to gain better understanding of the use of pesticides for disease and pest management and their impacts on human health and biodiversity within two agricultural production landscapes in Ghana. Specifically, the study sought to:

1. Determine farmers' level of dependency on pesticides in managing diseases and pests on farms.
2. Estimate the extent of use of unapproved pesticides on cocoa farms.
3. Assess farmers' perceptions of the impacts of pesticides on the health of humans and the environment.
4. Evaluate the effects and benefits of alternatives (IPM and organic pesticides) on protected areas and the environment.
5. Document any COVID-19 and gender-related experiences relevant to the biodiversity-health-sustainability nexus.

2 Methodology

2.1 Study Area

The study was conducted within two major agricultural production landscapes in Ghana. These are the Bia Conservation Area (BCA) and Kakum Conservation Area (KCA) in the Western North and Central Regions of Ghana, respectively. The BCA includes the Bia, Enchi, Juaboso, and Sefwi Wiawso districts. The KCA includes the Assin North, Assin South, Lower Denkyira, and Upper Denkyira districts (Fig. 12.1 and Table 12.1). The two landscapes cover a total area of 34, 987 km^2 and fall within the Upper Guinean forest hotspot, one of the most biodiversity-rich hotspots of Ghana. The landscapes also contain 75% of primary forest and lie in the equatorial climatic zone, which is characterised by moderate temperatures. The landscapes lie in the moist deciduous forest ecosystem part of Ghana, with an average rainfall of 1600 mm per annum. They are characterised by a number of isolated forest patches that contain exceptionally diverse ecological communities, distinctive biodiversity, and a mosaic of forest types that provides refuge to numerous endemic species (Conservation Alliance (CA), 2018). The cultivation of cocoa and oil palm constitutes the main commodity-driven deforestation within the landscape. Agriculture is the main economic activity within both landscapes, dominated by tree and food crops. Cocoa is a major tree crop accounting for more than 55% of Ghana's tree crop exports (Kolavalli & Vigneri, 2011; Denkyirah et al., 2016). The BCA and KCA accounted for more than half (56%) of cocoa production in Ghana in 2019 (COCOBOD, 2020). The favourable edaphic conditions and the rich ecosystem largely account for the high production of cocoa.

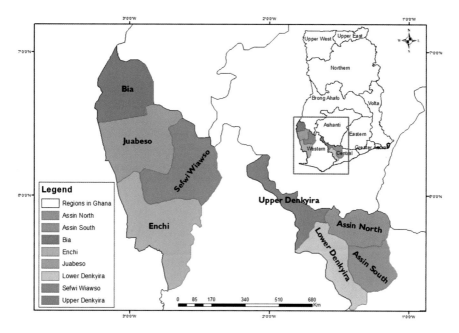

Fig. 12.1 Map showing the study landscapes: BCA (left) and KCA (right) (source: Conservation Alliance (CA), 2020)

Table 12.1 Basic information of the study area (source: Conservation Alliance (CA), 2020)

Country	Ghana
Province	Central and Western North Regions
Districts	Bia, Enchi, Juaboso, Sefwi Wiawso, Assin North, Assin South, Lower Denkyira, Upper Denkyira
Municipality	n.a.
Size of geographical area (hectare)	3,498,700
Number of direct beneficiaries (persons)	1500
Number of indirect beneficiaries (persons)	5000
Dominant ethnicity(ies), if appropriate	*Akans*
Size of the case study/project area (hectare)	780,400
Geographic coordinates (latitude, longitude)	6°45′35.1″N, 3°04′44.3″W

2.2   Study Population and Sampling Method

The total number of registered cocoa farmers is about 1500 in the Western North and Central regions (900 and 600 members, respectively) (Table 12.1). Among registered members, 70% are men and 30% women. A simple random sampling technique was adopted in selecting members of the Cocoa Conservation Association (CCA) for the study. A sample size of 306 was derived from the random selection of respondents of which 184 were from Western North Region and 122 from the Central Region based on the respective population size (Kirk, 2011; Etikan & Bala, 2017). The sample sizes were further divided to get an equal representation of the gender groupings. For the BCA, 129 male and 55 female farmers were sampled. Similarly, 85 male and 37 female cocoa farmers were sampled in the Central Region within the KCA. Images of interviewed respondents are shown in Fig. 12.2.

2.3   Data Collection

Both quantitative (survey) and qualitative research approaches were used. Farmer surveys, community consultations, focus group discussions, district-wide

Fig. 12.2 Researcher interviewing input seller (**a**), CCA members (**b**), and individual farmers (**c** and **d**) (Source: Conservation Alliance (CA), 2020. Photo credit: All images taken by CA Communications Team)

stakeholder consultations, and key informant interviews constituted the main strategies to generate data. Both primary and secondary data constituted a useful set of data that defined the state of pesticide usage and impacts among cocoa farmers within the landscape.

Surveys were conducted with farmers of the sample to generate primary data. Secondary data was secured from various sources including public records, institutional documents, management plans, policy statements, official publications, and other research works. The questionnaires for the survey were administered in the various districts in September 2020. Reviews of government documents on the CODAPEC programme, including operational manuals and monitoring reports, constituted another important form of data collection.

A series of key interviews were also conducted with the FSD, WD, EPA, and Protected Area Managers to assess the effects of alternatives to pesticide use on the health of protected areas. With respect to the effects of unapproved pesticides on human health, the study team conducted a risk assessment based on farmers' experiences in relation to the likelihood of occurrence and consequences of use of the unapproved pesticides.

2.4   Data Analysis

The questionnaires for the survey of farmers were pretested using a selected group of respondents outside the target population to ensure the reliability and validity of the questions and responses. All data were coded, and analysis was carried out using Statistical Package for Social Sciences (SPSS) and Microsoft Excel. The data obtained reflected the views, opinions, and attitudes of the respondents and further enhanced the reliability, validity, credibility, and accuracy of the results. The data obtained are summarised in tables for simplicity of analysis. The analysis was done using descriptive analysis where issues of similarity and dissimilarities of responses were compared. The descriptive statistical tools facilitated the quantitative comparative analysis of the responses.

3   Results

3.1   Demographic Characteristics of Respondents

The key demographic characteristics that were generated during the study included the age, years of experience, level of education, and marital status of respondents. Almost 70% of the respondents were male, arising from the dominance of males in the sector (Table 12.2). Additionally, over 77% of the respondents were over 41 years of age and almost 80% of the respondents had more than 10 years of work experience. Education plays a significant role in the acquisition and use of new

Table 12.2 Demographic characteristics of farmers (source: Conservation Alliance (CA), 2020)

Variable	Category	Number of Respondents	Percent (%)
Gender	Male	214	69.9
	Female	92	30.1
Age (Years)	<30	5	1.5
	30-40	65	21.3
	41-50	90	29.3
	51-60	81	26.5
	61-70	47	15.3
	>70	18	6.0
Marital Status	Married	256	83.6
	Divorced	17	5.7
	Single	33	10.7
Education	Non-Formal	130	42.5
	Basic	96	31.3
	Secondary	73	24.0
	Tertiary	7	2.2
Years of Experience	1 – 5	15	5.0
	6 – 10	47	15.3
	> 10	244	79.7

Fig. 12.3 Use of pesticides (source: Conservation Alliance (CA), 2020)

Use of Pesticides

18.7%

81.3%

■ Yes ■ No

technology. With more than 40% of respondents having no formal education, the risk of unapproved pesticide use is high.

3.2 Farmers' Level of Dependency on Pesticides

The use of pesticides is widespread among cocoa farmers due primarily to the high level of disease and pests and the ready supply of free inputs. The study revealed that about 81.3% of farmers are solely dependent on pesticides and consider them as the most important method for managing diseases and pests (Figs. 12.3 and 12.4). About 18.7% of the respondents are reliant on IPM, including cultural methods to control diseases and pests on farms, and depend less on pesticides.

Fig. 12.4 Farmer spraying (source: Conservation Alliance (CA), 2020. Photo: CA Communications Team)

3.3   Use of Unapproved Pesticides

The study revealed that at least 30% of the sampled farmers have continuously applied unapproved pesticides on their farms. *Akate Suro* (Diazinon) and *Consider* (Imidacloprid) were the most widely used because they are readily available on the market and their price is affordable (Table 12.3). *Carbamult* (Promecarb) and *So Bi Hwe* (a suspected cocktail of several active ingredients), on the other hand, were considered to be the most effective in controlling diseases and pests. The rest of the farmers have consistently benefited from the government's free supply of approved pesticides.

The risk matrix (Table 12.4) revealed that *Sumitox* (chlorpyrifos), *Akate Suro* (diazinon), *So Bi Hwe* (suspected cocktail), and *Lambda Super* (lambda-cyhalothrin) had the most devastating impacts on human health. The study recorded 23 reported cases of pesticide poisoning among cocoa farmers within the study areas within the past 5 years. The BCA, which accounts for almost half (45%) of cocoa output in Ghana, coincidentally also recorded about 78% of the reported cases, and the KCA (11% of cocoa output) accounted for about 22% of reported cases of human health impacts (Table 12.5).

Table 12.3 Unapproved pesticides used on cocoa by farmers (source: Conservation Alliance (CA), 2020)

Pesticide	Active ingredient	Importing companies	Exporting countries
Carbamult	Promecarb	Smuggled	Cote d'Ivoire
Sumitox	Chlorpyrifos	Kumark Ghana Limited	India
Consider	Imidacloprid	Thomas Fosu Enterprise	China
Akate Suro	Diazinon	Mybarnes Limited	Israel
Lambda Super	Lambda-Cyhalothrin	Kumark Ghana Limited	China
So Bi Hwe	Mixture of two or more of the above active ingredients	Smuggled	Cote d'Ivoire

Table 12.4 Human health risk matrix of unapproved pesticides (source: Conservation Alliance (CA), 2020)

	Health Consequences					
Likelihood	**Skin Irritation**	**Eye Irritation**	**Dizziness**	**Breathlessness**	**Headache**	**Vomiting**
Almost Certain	*So Bi Hwe* (Mixture of two or more active ingredients)	*Lambda Super* (Lambda-Cyhalothrin) *Akate Suro* (Diazinon)	*Sumitox* (Chlorpyrifos) *Lambda Super* (Lambda-Cyhalothrin)	*Lambda Super* (Lambda-Cyhalothrin) *Akate Suro* (Diazinon)	*Sumitox* (Chlorpyrifos) *Akate Suro* (Diazinon)	*Sumitox* (Chlorpyrifos) *Akate Suro* (Diazinon) *Lambda Super* (Lambda-Cyhalothrin)
Likely	*Lambda Super* (Lambda-Cyhalothrin)	*Sumitox* (Chlorpyrifos)	*Carbamult* (Promecarb)	*So Bi Hwe* (Mixture of two or more active ingredients)	*Carbamult* (Promecarb)	*Carbamult* (Promecarb)
Possible	*Akate Suro* (Diazinon)	*So Bi Hwe* (Mixture of two or more active ingredients)	*Consider* (Imidacloprid)	*Consider* (Imidacloprid)	*So Bi Hwe* (Mixture of two or more active ingredients)	*So Bi Hwe* (Mixture of two or more active ingredients)

Notes: The probability of health effects occurring: Almost Certain (100%), Likely (50%), Possible (≤25%)

Consequence is **very critical** requiring immediate medical attention.
Consequence is **major** requiring specific treatment to contain it.
Consequence is **moderate** requiring normal routine management and effects fade after sometime.

3.4 Farmer's Perception of the Impacts of Pesticides

The use of pesticides is perceived to adversely affect the health of humans. While over 90% of the respondents were aware of the adverse impacts of pesticides on humans because of the visible signs they leave behind, only a few (less than 10%) knew of the effect on the environment (biodiversity) (Fig. 12.5).

During the study, the participating farmers shared their experiences related to the prolonged use of pesticides in cocoa production (Fig. 12.6) and reported on symptoms experienced. Regarding the effects on human health, the farmers reported skin

Table 12.5 Reported cases of pesticide poisoning (source: Conservation Alliance (CA), 2020)

Production landscape	Districts	No. of reported cases	Percentage of reported cases
Bia conservation area	Bia	8	34.8
	Enchi	3	13.0
	Sefwi Wiawso	3	13.0
	Juaboso	4	17.4
Kakum conservation area	Assin North	2	8.7
	Assin South	1	4.3
	Lower Denkyira	1	4.3
	Upper Denkyira	1	4.3
Total		**23**	**100.0**

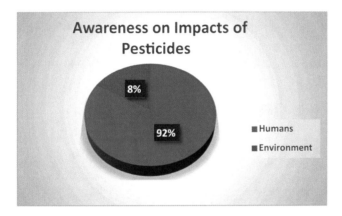

Fig. 12.5   Awareness on the impacts of pesticides (source: Conservation Alliance (CA), 2020)

irritation (more than 30%) to be the most commonly experienced effect on human health. Others included eye irritation (21%), dizziness (18%), vomiting (16%), and breathlessness (13%). About 5% of the farmers appeared indifferent to the adverse effects of pesticides and argued that such effects on humans could only arise from the state of the individual's health, particularly the toughness of the skin.

With respect to the less than 10% of respondents who reported impacts on biodiversity, about half of the reported impacts of pesticide use were perceived to cause the death of vertebrates (e.g. rodents, reptiles) as shown in Fig. 12.7. About 35% of reported cases involved harm to a wide range of beneficial invertebrate species (e.g. earthworms in the soil and other terrestrial habitats), and pollinators (e.g. bees, midges, and butterflies). About 15% of pesticide impacts reported were related to adverse effects on freshwater aquatic species (e.g. snails and water fleas), which were affected via water and aquatic plants exposed to pesticide residue.

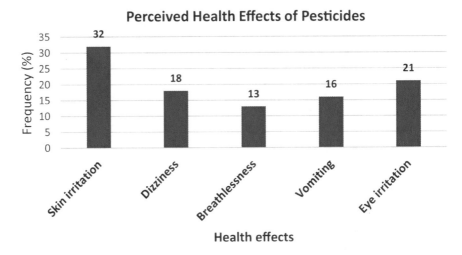

Fig. 12.6 Perceived health effects of pesticides (source: Conservation Alliance (CA), 2020)

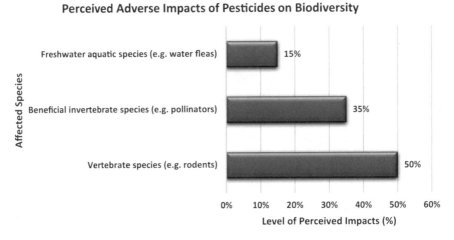

Fig. 12.7 Perceived impacts of pesticides on biodiversity (source: Conservation Alliance (CA), 2020)

3.5 Effects and Benefits of Alternatives (IPM and Organic Pesticides) on Protected Areas and the Environment

With respect to alternatives to conventional pesticides used by farmers that are ecologically friendly to protected areas and the environment, the study recorded about a 69.8% patronage of IPM and 30.2% of biopesticides (organic pesticides). The most commonly used biopesticides were neem tree (*Azadirachta indica*) extract (55%); pyrethrum 5EW, a natural substance purified from chrysanthemum flowers

Table 12.6 Benefits of alternatives (IPM and organic pesticides) (source: Conservation Alliance (CA), 2020)

Benefits	No. of recorded benefits	% of recorded benefits
Reduced environmental risk associated with pest management by encouraging the adoption of more ecologically benign control tactics	17	18
Protection of non-target species including flora and fauna	22	23
Reduced air and groundwater contamination	10	10
Alleviation of public concern about pests and pesticide-related practices	13	13
Promotes sustainable bio-based pest management alternatives, thus reducing the need for pesticides	15	16
Maintains or increases the cost-effectiveness of a pest management programme	19	20
Total	**96**	**100**

(31%); and milk bush (*Thevetia peruviana*) extract (14%). Additionally, 86% of the respondents attested to IPM and biopesticides being effective in pest control and protection of non-target flora and fauna. The study also assessed the benefits of adopting other alternatives (IPM and organic pesticide) and recorded 96 benefits from the landscape through monitoring reports of the FSD, WD, and EPA (Table 12.6).

According to the study, SEPL management by the community through sustainable practices such as IPM and biopesticide application had positive beneficial effects contributing to the well-being of humans and environment, as shown in Fig. 12.8. Therefore, this interdependent relationship among the three sustainability pillars through IPM and biopesticides offers prospects for local communities and farmers to adopt practices compatible with living in harmony with nature to enhance the ecological health of their landscapes.

3.6   Gender and COVID-19-Related Issues

The study also documented some gender and COVID-19-related issues in the landscapes, which were relevant to the biodiversity-health-sustainability nexus. The study revealed that 60% of male respondents and 58% of female respondents were aware that some pesticides were unapproved by COCOBOD. However, only 5% of the female respondents were able to specify these unapproved pesticides, as compared to 65% of the male respondents. Additionally, the study examined how the COVID-19 pandemic has impacted the purchasing power of males and females with respect to the use of approved and unapproved pesticides. Since the COVID-19 pandemic has negatively impacted the income of farmers, use of pesticides is based on affordability. The study revealed that male cocoa farmers (65%) were able to

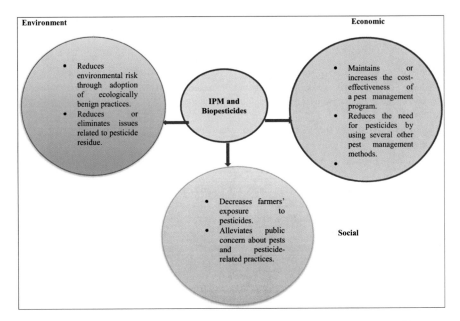

Fig. 12.8 Community conceptualisation of benefits of IPM and biopesticides (source: Conservation Alliance (CA), 2020)

afford approved pesticides more than their female counterparts (25%), since unapproved pesticides were found to be less expensive than the approved ones.

4   Discussion

The study found there to be an over-reliance on pesticides by cocoa farmers in addressing incidences of disease and pests on farms. This is to be expected considering that these pesticides are freely distributed to farmers to help increase national cocoa output. While the government programme has resulted in an increase in the national cocoa output from about 600,000 MT to 1,000,000 MT, it also negatively impacted the health of humans and the environment (Conservation Alliance (CA), 2020).

Some farmers were still reliant on unapproved pesticides, which are mostly HHPs. Even though the law forbids the importation, marketing, and use of unapproved pesticides, this regulation is not sufficiently enforced and the penalty is not enough of a deterrent to stop illegal practices, largely accounting for their widespread sale and use. Use of unapproved pesticides correlates with low purchasing power, perception of the poor effectiveness of approved pesticides, and ignorance about which pesticides are approved, especially in developing countries (Denkyirah et al., 2016). The Transnational Alliance to Combat Illicit Trade

(TRACIT) confirms that the use of these types of pesticides makes enforcement of regulations more difficult, while also increasing the danger to human health and the environment (Transnational Alliance to Combat Illicit Trade (TRACIT), 2019).

As observed by the farmers, pesticides cause prominent health and environmental impacts, having adverse effects on pollinators, soil organisms, and human health (e.g. acute headaches and breathlessness) (Özkara et al., 2016). In particular, pesticides cause significant damage and pose serious risks to a wide range of beneficial invertebrate species in the soil, vegetation, and aquatic habitats (Bernardes et al., 2015). These organisms play vital roles in ecosystem functioning and services, including pollination, regulating of organic wastes, nutrient cycling, food production, biological pest control, and water cleansing (Power, 2010). The most damaging ecological disturbances of excessive pesticide use include the existence of high concentration of pesticide residues in food chains and biodiversity loss (Boakye, 2012). A study by Dankyi (2015) found that neonicotinoids are persistent in Ghanaian soils even after several months to years of application and have a high potential of being leached into surface and underground water systems. According to a study by Pesticide Action Network (PAN) UK (2018), neonicotinoid pesticides in cocoa production/farming can present acute health effects. According to the PAN UK study, US farm workers have similarly reported skin or eye irritation, dizziness, breathlessness, confusion, or vomiting after exposure to HHP pesticides (e.g. imidacloprid). The same study by PAN UK highlighted serious harm to a wider range of organisms at the levels which threaten the essential ecosystem services of pollination, nutrient cycling, and natural pest control.

This study assessed alternatives to HHPs and showed that the majority of farmers favour IPM (70%) as potential alternative and an effective means of pest and disease control and maintenance of the socio-ecological health of the landscape. This was reinforced by the number of recorded benefits of IPM, including protection of biodiversity. According to Mboussi et al. (2018), IPM thus represents a better option for sustaining cocoa yield and enhancing the ecological health of landscapes in view of the multiple interrelated practices. A similar study by Dara (2019) showed that the adoption of IPM strategies and organic pesticides provides economic benefits by reducing pest damage to crops, thus increasing productivity. According to Pesticide Action Network (PAN) UK (2018), the adoption of IPM enhances benefits for biodiversity, the environment, and food productivity as compared to the use of synthetic (chemical) pesticides. IPM has greatly reduced pesticide poisoning incidences within the production landscape. Additionally, some community members integrate tree cultivation into their agricultural lands not only to reduce disease and pest incidences, but also to enhance the ecological health of the farmlands. As a result, 12 cocoa farms belonging to some cocoa cooperatives have been turned into demonstration plots for promoting best practices in pest and disease management within the project landscape. Additionally, the increased awareness on the evidence of ecological and health benefits of IPM and biopesticides has resulted in increased adoption among community members.

The study revealed that there is a low level of knowledge among women concerning the illegal practice of using unapproved pesticides, which are mostly

HHPs. This can be attributed to low education, less participation during sensitisation exercises, inability to read, and lack of involvement during group meetings, thus exposing women to greater risk from highly hazardous pesticides as compared to men. According to Kawarazuka et al. (2020), despite gender differences, research and training on pest and disease management often target farmers as a whole neglecting the specific needs of women and men. This study also revealed that the lower financial status of women is a key driver for women to use unapproved pesticides, which were mostly highly hazardous. Poverty is one of the key reasons why the envisioned sustainable cocoa sector is failing. The same poverty has compelled cocoa farmers, especially women farmers who are not able to purchase approved pesticides, to resort to cheaper but more hazardous pesticides especially in the face of the COVID-19 pandemic (Walker, 2021). Additionally, the high percentage of males in cocoa farming is likely linked with the involvement of the women in the cultivation of other crops and trading of food crops. Accordingly, most of the training programmes organised by COCOBOD have often targeted male cocoa farmers with the participation of very few women (Williamson, 2011).

5 Conclusion

This study sought to document the various methods commonly used by smallholder cocoa farmers to manage pests and disease incidence, and to establish their level of dependency on pesticides. The dependence on pesticides by farmers has adversely affected the socio-ecological health of the production landscapes. Governments are advised to take steps to phase out active ingredients of HHP pesticides from the pesticide value chain due to their adverse effects on human health and the environment. Regulatory agencies must be further empowered to regulate these products on the market and clamp down on unapproved pesticides with very deterrent penalties including lengthy imprisonments, huge fines, and suspension of operations of companies and individuals that disobey the laws. Additionally, policymakers are encouraged to promote the use of environmentally friendly pesticides, including biopesticides, which are relatively safe to humans and biodiversity. Steps must be taken to stimulate investments in the production and distribution of biopesticides so that they are available, accessible, and affordable to cocoa farmers. Furthermore, there must be intensification of farmer education and training on IPM coupled with intensified supervision and monitoring by extension officers to encourage the increased adoption of IPM to enhance the socio-ecological health of the cocoa landscapes in Ghana.

References

Adjinah, K. O., Opoku, I. Y. (2010). The National Cocoa Diseases and Pest Control (CODAPEC): Achievement and Challenges, Ghana Cocoa Board, viewed 30 February 2020. Retrieved from News.Myjoyonline.Com/Features/201004/45375.Asp.

Afrane, G., & Ntiamoah, A. (2011). Use of pesticides in the cocoa industry and their impact on the environment and the food chain. In D. M. Stoytcheva (Ed.), *Pesticides in the modern world—risks and benefits* (pp. 51–68). IntechOpen, viewed 23 June 2021. Retrieved from http://www.intechopen.com/books/pesticides-in-the-modern-world-risks-and-benefits/use-of-pesticides-in-thecocoa-industry-and-their-impact-on-the-environment-and-the-food-chain.

Amoah, S. K. (2013). Factors affecting cocoa production in Upper Denkyira West District. M.Sc. Dissertation, Kwame Nkrumah University of Science and Technology.

Anang, B. T., Mensah, F., & Asamoah, A. (2013). Farmers' assessment of the government spraying program in Ghana. *Journal of Economics and Sustainable Development, 4*(7), 92–99.

Bernardes, M. F. F., Pazin, M., Pereira, L. C., & Dorta, D. J. (2015). Impact of pesticides on environmental and human health. In A. C. Andreazza (Ed.), *Toxicology studies—cells, drugs and environment*. IntechOpen, viewed 4 February 2021, Retrieved from https://www.intechopen.com/chapters/48406.

Boadu, M. O. (2014). Assessment of pesticides residue levels in cocoa beans from the Sefwi Wiawso District of the Western Region of Ghana. M.Sc. Dissertation, Kwame Nkrumah University of Science and Technology.

Boakye, S. (2012). Levels of selected pesticide residues in cocoa beans from Ashanti and Brong Ahafo Regions of Ghana. M.Sc. Dissertation, Kwame Nkrumah University of Science and Technology.

COCOBOD. (2020). *Cocoa Production Statistics*, Research Department, Ghana Cocoa Board (COCOBOD), Accra, viewed 5 May 2021. Retrieved from https://cocobod.gh/cocoa-purchases.

Conservation Alliance (CA). (2018). The state of biodiversity within cocoa production landscape in Southwest Ghana. *Cocoa Biodiversity, 5*, 50–150.

Conservation Alliance (CA). (2020). *Assessment on the Gender Dynamics of Highly Hazardous Pesticides (HHPs) within Cocoa Production Landscape in Ghana*, A research report for INKOTA Netzwerk and BMZ, Germany, viewed 19 October 2020. Retrieved from https://conservealliance.org/conservation-alliance-ca-releases-a-study-report-on-assessment-on-the-gender-dynamics-of-highly-hazardous-pesticides-hhps-within-cocoa-production-landscape-in-ghana/.

Conservation Alliance (CA). (2019). Farmers' management of disease and pest incidence within cocoa production landscape in Ghana. A research report for UTZ sector partnerships program GHANA.

Damalas, C. A., & Eleftherohorinos, I. G. (2011). Pesticide exposure, safety issues, and risk assessment indicators. *International Journal of Environmental Research and Public Health, 8*(5), 1402–1419.

Dankyi, E. (2015). Exposure and fate of neonicotinoid insecticides in cocoa plantations in Ghana. PhD dissertation, Department of Chemistry, University of Ghana.

Dara, S. K. (2019). The new integrated pest management paradigm for the modern age. *Journal of Integrated Pest Management, 10*(1), 12. https://doi.org/10.1093/jipm/pmz010

Denkyirah, E. K., Okoffo, E. D., Adu, D. T., Aziz, A. A., Ofori, A., & Denkyirah, E. K. (2016). Modeling Ghanaian cocoa farmers' decision to use pesticide and frequency of application: the case of Brong Ahafo Region. *SpringerPlus, 5*(1), 1–17, viewed 26 March 2021. https://doi.org/10.1186/s40064-016-2779-z

Essegbey, G. O., & Ofori-Gyamfi, E. (2012). Ghana Cocoa industry—an analysis from the innovation system perspective. *Technology and Investment, 3*, 276–286.

Etikan, I., & Bala, K. (2017). Sampling and sampling methods. *Biometrics & Biostatistics International Journal, 5*(6), 215–217. https://doi.org/10.15406/bbij.2017.05.00149

Gakpo, J. O. (2012). Nurturing a youthful generation of cocoa farmers. *Modern Ghana*, Feature Article, 30 May, viewed 20 June 2021. Retrieved from http://www.modernghana.com/news/3 98735/1/nurturing-a-youthful-generation-of-cocoa-farmers.html.

Gill, H. K., & Garg, H. (2014). Pesticides: Environmental impacts and management strategies. In S. Soloneski (Ed.), *Pesticides—toxic aspects*. IntechOpen. https://doi.org/10.5772/57399

ISSER. (2011). *The State of the Ghanaian Economy in 2010*. The Institute of Statistical Social and Economic Research, University of Ghana.

Kirk, R. E. (2011). Simple random sample. In M. Lovric (Ed.), *International encyclopedia of statistical science*. Springer. https://doi.org/10.1007/978-3-642-04898-2_518

Kolavalli, S., & Vigneri, M. (2011). Cocoa in Ghana: Shaping the success of an economy. In P. Chuhan-Pole & M. Angwafo (Eds.), *Yes Africa can: success stories from a dynamic continent* (pp. 201–217). The World Bank. https://doi.org/10.1596/978-0-8213-8745-0

Kawarazuka, N., Damtew, E., Mayanja, S., Okonya, J. S., Rietveld, A., Slavchevska, V., & Teeken, B. (2020). A gender perspective on pest and disease management from the cases of roots, tubers and banana in Asia and sub-Saharan Africa. *Frontiers in Agronomy, 2*, 7. https://doi.org/10.3389/fagro.2020.0007

Mboussi, S. B., Ambang, Z., Kakam, S., & Beilhe, L. B. (2018). Control of cocoa mirids using aqueous extracts of Thevetia peruviana and Azadirachta indica. *Cogent Food & Agriculture, 4*, 1430470. https://doi.org/10.1080/23311932.2018.1430470

Özkara, A., Akyil, D., & Konuk, M. (2016). Pesticides, environmental pollution, and health. In M. L. Larramendy (Ed.), *Environmental health risk—hazardous factors to living species*. IntechOpen, viewed 23 July 2021, https://www.intechopen.com/chapters/50482.. https://doi.org/10.5772/63094

Pesticide Action Network (PAN) UK. (2018). *Pesticide Use in Ghana's Cocoa Sector: Key findings*. A consultancy report for UTZ Sector Partnerships program GHANA. https://utz.org/wp-content/uploads/2018/06/18-05-Key-Findings-Report-on-Pesticide-Use-in-Ghana.pdf.

Power, A. G. (2010). Ecosystem services and agriculture: trade-offs and synergies. *Philosophical Transactions of the Royal B Biological Sciences, 365*(1554), 2959–2971. https://doi.org/10.1098/rstb.2010.0143

Singh, A. U., & Prasad, D. (2016). Integrated pest management with reference to INM. *Advances in Crop Science and Technology, 4*, 220. https://doi.org/10.4172/2329-8863.1000220

Transnational Alliance to Combat Illicit Trade (TRACIT). (2019). Chapter 2. SDGs and illicit trade in agrochemicals and pesticides. In*Mapping the impact of illicit trade on the sustainable development goals* (pp. 20–30). Transnational Alliance to Combat Illicit Trade.

Vigneri, M., & Kolavalli, S. (2018). *Growth through pricing policy: the case of cocoa in Ghana*. Background paper to the UNCTAD-FAO Commodities and Development Report 2017. Food and Agriculture Organization.

Williamson, S. (2011). Understanding the full costs of pesticides: Experience from the field, with a focus on Africa. In M. Stoytcheva (Ed.), *Pesticides—The impacts of pesticides exposure*. IntechOpen. https://doi.org/10.5772/14055

World Bank Group. (2018). *3rd Ghana economic update: Agriculture as an engine of growth and jobs creation, African Region*, viewed 17 May 2021. Retrieved from https://documents1.worldbank.org/curated/pt/113921519661644757/pdf/123707-REVISED-Ghana-Economic-Update-3-13-18-web.pdf.

Walker, J. (2021). Fairtrade urges EU to back living incomes in West African cocoa supply chains. Confectionary Production news, viewed 12 April 2021. Retrieved from https://www.confectioneryproduction.com/news/34536/fairtrade-urges-eu-to-back-living-incomes-in-west-african-cocoa-supply-chains/.

The opinions expressed in this chapter are those of the author(s) and do not necessarily reflect the views of UNU-IAS, its Board of Directors, or the countries they represent.

Open Access   This chapter is licenced under the terms of the Creative Commons Attribution 3.0 IGO Licence (http://creativecommons.org/licenses/by/3.0/igo/), which permits use, sharing, adaptation, distribution and reproduction in any medium or format, as long as you give appropriate credit to UNU-IAS, provide a link to the Creative Commons licence and indicate if changes were made.

The use of the UNU-IAS name and logo, shall be subject to a separate written licence agreement between UNU-IAS and the user and is not authorised as part of this CC BY 3.0 IGO licence. Note that the link provided above includes additional terms and conditions of the licence.

The images or other third party material in this chapter are included in the chapter's Creative Commons licence, unless indicated otherwise in a credit line to the material. If material is not included in the chapter's Creative Commons licence and your intended use is not permitted by statutory regulation or exceeds the permitted use, you will need to obtain permission directly from the copyright holder.

Chapter 13
Synthesis: Concept, Methodologies, and Strategies to Address the Nexus in SEPLS

Maiko Nishi, Suneetha M. Subramanian, and Himangana Gupta

Abstract This chapter synthesises major findings from the eleven case studies presented in the previous chapters, offering policy recommendations arising from the synthesis. It distills key messages to address questions on the following issues: (1) how to conceptualise the nexus between biodiversity, health, and sustainable development in the context of SEPLS management; (2) how to measure, evaluate, and monitor the effectiveness of SEPLS management in regard to securing and improving both ecosystem and human health; and (3) how to address the challenges and seize the opportunities of SEPLS management in minimising trade-offs and maximising synergies between different efforts augmenting both ecosystem and human health, as well as well-being, so as to move towards more sustainable futures. The chapter identifies several policy recommendations to better manage the biodiversity-health-sustainability nexus in SEPLS and facilitate transformative change for sustainable development. It also revisits the concept of the biodiversity-health-sustainability nexus to offer perspectives on the complex interlinkages in the context of managing SEPLS on the ground.

Keywords Socio-ecological production landscapes and seascapes · Nexus · Trade-offs · Synergies · One Health · Integrated approaches · Sustainable development

Contributing authors to this chapter include Nadia Bergamini, Valerie Braun, Jung-Tai Chao, Dipayan Dey, Andrea Fischer, Chris Jacobson, Paulina G. Karimova, N. Anil Kumar, Kuang-Chung Lee, María Elena Méndez López, Patrick Maundu, Yasuyuki Morimoto, William Olupot, Yaw Osei-Owusu, Raymond Owusu-Achiaw, Md. Shah Paran, Vipindas P, Andrés Quintero-Ángel, Sara Catalina Rodriguez Dias, Jeeranuch Sakkhamduang, Fausto O. Sarmiento, V.V. Sivan, and Rashed Al Mahmud Titumir.

M. Nishi (✉) · S. M. Subramanian · H. Gupta
United Nations University Institute for the Advanced Study of Sustainability (UNU-IAS), Tokyo, Japan
e-mail: nishi@unu.edu

© The Author(s) 2022

267

M. Nishi et al. (eds.), *Biodiversity-Health-Sustainability Nexus in Socio-Ecological Production Landscapes and Seascapes (SEPLS)*, Satoyama Initiative Thematic Review, https://doi.org/10.1007/978-981-16-9893-4_13

1 How Can We Conceptualise the Nexus Between Biodiversity, Health, and Sustainable Development in the Context of SEPLS Management?

The **nexus between biodiversity, health, and sustainable development** can be seen as the interdependent connections between the ecological and human dimensions of a social-ecological system that together contribute to the well-being and resilience of the human-nature system (see Chap. 1). Such a conceptualisation furthers a recognition of the links between biodiversity and health in the sense of physical, mental, and spiritual health. Furthermore, human health and well-being are dependent on environmental health that provides various ecosystem benefits we derive including clean air, water, regulation of floods, food security, nutritional security, medicinal resource diversity, and access to various cultural benefits, among others. The understanding of this dependence of human health on the environment has become more vivid during the COVID-19 outbreak, when multiple challenges arising from the viral disease, including those relating to access to medical facilities, conventional food markets, employment opportunities, and the like, came to the fore especially in the context of local communities. We examine in this volume (and specifically this chapter) how communities cope and find solutions to address these multiple challenges within the context of socio-ecological production landscapes and seascapes (SEPLS). Specifically in the context of SEPLS, biodiversity acts like insurance to ensure security of food, health, energy, water, livelihoods, and income, among others. This is because a diversity of resources (and by implication well-functioning ecosystems) enhance options to mitigate or adapt to various perturbances (natural or human-induced) through regulating natural processes (such as floods and soil fertility), providing varieties that can adapt to climatic change, and opening economic opportunities that allow diversifying risks from economic or natural shocks, leading to better health outcomes. It is important to note that securing these benefits is dependent on the wise use of both traditional and modern knowledge relating to use, maintenance, and marketing of products and services from the SEPLS.

1.1 Human Health and Healthy Environment

Health and Sustainable Development

Health is defined as physical, mental, and social well-being (WHO, 2006), implying a deep and wide correlation that connects the environment with social relations and economic activities. What this means is that ensuring good health is a pathway towards achieving the Sustainable Development Goals (SDGs). For example, unravelling the various constituents of good health at the local level has been well highlighted in connection to food security and sovereignty brought about by dietary diversity (for instance, see Chaps. 9 and 11). Also well understood are access to medicinal resources and spaces of cultural and aesthetic significance that are also important for mental well-being (in addition to physical health). This implies that ensuring good health at the community level strongly contributes to a high level of social-ecological resilience. In this context, the following components, in no specific order, have been identified as relevant parts of the nexus between biodiversity, health, and sustainable development:

- *Ensuring food security and sovereignty,*[1] *dietary diversity, and nutritional security:* To achieve this, ensuring a high degree of agro-biodiversity, furthering healthy soils, and "no-harm" agronomic practices (such as rational or low use of chemical pesticides and increase in ecologically friendly organic pesticides) are required. These have consequent benefits for the health of water systems (including coastal and marine ecosystems), other biological life forms, and human health (e.g. Chaps. 3–5, 10–12).
- *Ensuring the integrity of ecosystems especially relating to the regulation of natural cycles such as of water, soil fertility, carbon, nitrogen, and phosphorus:* Achieving this requires management of water, plant, and soil resources in a sustainable fashion and may involve regeneration activities. These activities have a direct impact on agricultural productivity, availability of clean air and water, as well as psychological health, in addition to enhancing biocultural linkages (e.g. Chaps. 4, 6, 7, and 11).
- *Ensuring One Health that includes the health of people, environment, and animals:* This requires interdisciplinary expertise from modern sciences and local knowledge to establish good practices for the integrated management of the health-for-all-linked components of human society and nature. This would require negotiating prejudices between disciplines, knowledge systems, and power hierarchies among different decision-making institutions. This nexus approach would enable policy coherence and a strong alignment of local-level priorities and policy goals on ensuring health and sustainability using a systems approach (e.g. Chaps. 2, 4, 9, and 12).
- *Ensuring that different stakeholders in the community are engaged proactively in the decision-making and management of SEPLS:* This ensures inclusion of various priorities related to well-being and bridges potential areas of conflict that could arise due to decisions related to the individual use and management of landscapes and seascapes. This would require an explicit acknowledgement of the rights and obligations of different stakeholders to the landscape and resources, including both primary and secondary stakeholders (e.g. Chaps. 4, 8, 10, and 12).
- *Ensuring connectivity between adjoining landscapes under different governance regimes:* This requires an explicit recognition of the heterogeneity of ecosystems that is to be factored in to planning and governance systems in landscapes and seascapes, including national and regional government reserves, production

---

[1] Sovereignty refers to control over types and supply of food.

areas, private lands, indigenous reservations, and other effective area-based conservation measures (OECMs). This will ensure conservation of biodiversity and ecosystems and their sustainable use, and will further ensure that the norms for access and management are fair, just, and equitable to all stakeholders (e.g. Chaps. 3, 4, and 5).

In summary, the nexus at the SEPLS level constitutes a combination of access to food, medicines, various ecosystem types and benefits, autonomy and co-management, and cultural and political rights.

1.2 What Does the Nexus Approach Mean in the SEPLS Context?

Based on the priorities identified in the nexus between biodiversity, health, and sustainability, we can further identify some clear scalar level guidance to the management and governance of SEPLS to strengthen these interconnections. These include the following:

- At the resource level:

  - Focus on promoting spatial connectivity (within and between landscape and seascape) that is crucial to the linkages between production practices and impacts on biodiversity and health of humans and other life forms.
  - Given the direct dependence of populations in SEPLS on biodiversity, for example for food, health, and nutrition, promote activities that enhance diversity of species and improve mediums that support life (such as soil, water), and encourage cultural practices that promote this diversity (e.g. endemic food, local food concepts, traditional medicine practices, sustainable production and consumption that are linked to cultural knowledge).
  - Enhance livelihood security based on the connectivity of traditional knowledge and modern sustainable customary practices such as aquaculture, small-scale fishing, crab culture, honey and wax culture, and traditional arts and crafts, as both alternative and additional livelihood options.

- At the governance level:

  - Focus on promoting inclusive planning among different stakeholder groups including women, youth, and other marginalised groups.
  - Promote cooperation and strategic alliances between public, private, and academic sectors with convergent goals and co-management scenarios.
  - Integrate traditional and modern knowledge for endogenous development, which works on the principle that all available knowledge and experience are leveraged to address well-being priorities on the ground for sustainable, regenerative development.

 – Promote intergenerational transfer and documentation of traditional knowledge and sustainable customary practices and sharing of experiences with
   other SEPLS and a wider set of communities that would foster relevant
   innovations.

- At the macro policy decision-making level:

 – Focus on initiatives that promote sustainability including "green" energy use;
   address issues of biodiversity loss, climate change, and livelihood security
   together; promote adaptive capacity; and reduce vulnerabilities to natural and
   socio-economic shocks (such as the pandemic, and other disasters from natural
   and economic or policy shocks).
 – Promote community-based approaches in national policymaking to encourage
   local solutions within larger Global North-Global South scenarios in obser-
   vance of the United Nations Decade on Ecosystem Restoration.
 – Expand recognition of communities' rights, appreciation of their knowledge
   and values, and protection of their lands in relevant policy processes at
   international, regional, and national levels with a particular focus on the
   Post-2020 Global Biodiversity Framework (GBF) phase.

2   Measuring, Evaluating, and Monitoring the Effectiveness of SEPLS Management with Regard to Ecosystem and Human Health

The health-biodiversity-sustainability nexus is distinctive when it comes to SEPLS
because of inherent human-nature interactions and scalar linkages. Although conventional methods and approaches for project monitoring and evaluation (M&E) are
well known (Bours et al., 2014; GIZ, UNEP-WCMC, and FEBA, 2020), monitoring
implementation of the interventions, particularly from the nexus lens, demands more
tailored approaches for SEPLS, based on the specific conditions of the multiple
ecosystems they represent, including their sociocultural diversity and well-being
parameters. In this section, we discuss the approaches, tools, challenges, and opportunities with regard to monitoring and evaluating the effectiveness of interventions
in SEPLS. The approaches and tools captured here emerge from the experience of
the practitioners of on-the-ground projects and initiatives. Figure 13.1 shows the
main keywords emerging out of these case studies with reference to M&E.

2.1   Approaches for Monitoring the Biodiversity-Health-Sustainability Nexus in SEPLS

The previous volumes of SITR highlighted the need for M&E in SEPLS and various
M&E methods, focusing on SEPLS' multifaceted and multidimensional nature that

value networks
model communities term
governance **stakeholder** mapping
empowerment
disciplinary **localisation** capacity
resource **youth** aerial participatory
planning **citizen** **indigenous** **local** imageries
women **community** **level** gis survey
awareness **knowledge** **multi** long
network based science
ecological **monitoring** building
accounting innovation
documentation practices
traditional

Fig. 13.1 Top keywords on relevance and methods of M&E approaches in SEPLS presented in this book

makes the need for integrated approaches much more essential (UNU-IAS & IGES, 2019; Nishi et al., 2021). In this volume, we discuss below such integrated approaches used by and mentioned in various case studies, namely:

Community engagement approaches: These include operationalising traditional knowledge systems, and building capacities of the community through facilitation, training, and engagement in localising global, regional, or national indicators. Communities need to be given greater ownership of the interventions, including the situational enhancement of the role of indigenous peoples and local communities (IPLCs), women, youth, and other marginalised groups in monitoring and documentation based on localised indicators.

Mixed approaches using both quantitative and qualitative methods: Although community engagement is deemed important, it is also useful to do economic valuation of biodiversity and ecosystem services to have quantitative estimates. To effectively use mixed approaches, studies on human-nature interactions must include social analyses to obtain a holistic picture, including qualitative aspects perceived from the ground. This transdisciplinary effort will help combine biophysical and social knowledge.

Interdisciplinary and multidisciplinary approaches: These approaches integrate expertise across different levels and sectors. For this to happen, there is a need to develop indicators that are locally sensitive and comprehensible to all involved stakeholders while taking into consideration the diversity of sectors and metrics involved in human-nature linkages that may need evaluation in terms of ecosystem and human health. The different stakeholders include the public sector (various ministries including health, agriculture, nature conservation, and environment), private sector, and local community. In the SEPLS context, we may mention this as a "whole-of-the-society" approach or an "interagency" approach.

Table 13.1 Synthesis of various approaches and tools used and suggested by the case studies

Approaches	Methods and tools
• Community empowerment and engagement of youth and women	• Use of indigenous and local knowledge • Revitalisation of traditional practices • Monitoring and documentation by the community • Participatory planning
• Resource mapping	• Local survey • Ecosystem service valuation
• Remote sensing	• Use of aerial imageries • Use of GIS data available at other platforms
• Localisation of indicators	• Selection of locally adapted indicators • Engagement or creation of local self-help groups to generate awareness on global indicators
• Networking	• Long-term ecological network • Engagement of multiple stakeholders

2.2   Tools and Indicators for M&E

Implementation of the above-mentioned approaches requires appropriate and need-based tools and indicators well suited to the SEPLS context. While no one tool or indicator alone would suffice for M&E in SEPLS, there are various options tried and tested by practitioners. Table 13.1 shows the various approaches and tools used in the case studies presented in this book. Some of them include the following:

SEPLS resilience indicators toolkit: This toolkit is unique in terms of its suitability and flexibility for a SEPLS as it offers a set of indicators that are designed for monitoring both ecosystem and human health and their linkages, but can be flexibly chosen and adapted through local consultation to accord with local conditions and circumstances.

Use of other available indexes: Other indexes, if well suited to the SEPLS condition, can be used, such as the Livelihood Vulnerability Index (LVI),[2] Social Vulnerability Index,[3] and other tools that include biodiversity mapping. Some of these already include multiple sectors or can be tweaked appropriately to include more indicators related to the ecosystem and human health nexus.

Setting benchmarks: For very specific local conditions, it is better to set appropriate benchmarks for the evaluation of ecosystem and human health including qualitative and quantitative changes in biodiversity and air, water, and soil quality.

---

[2]The LVI uses multiple indicators to assess exposure to natural disasters and climate variability, social and economic characteristics of households that affect their adaptive capacity, and current health, food, and water resource characteristics that determine their sensitivity to climate change (Hahn et al., 2009).

[3]Developed by the Agency for Toxic Substances and Disease Registry (USA), it was meant to create databases to help emergency response by planners and public health officials to facilitate identification and mapping of communities that will most likely need support before, during, and after a hazardous event. It is increasingly used as a part of a climate resilience toolkit.

This can also help in long-term resource mapping including degradation in ecosystem or human health, and effectiveness of SEPLS management and conservation.

Adapting global indicators: Although several global indicators are available from the SDGs, biodiversity, and health targets, many of them may not be well suited for local conditions. Such indicators can be tweaked or downscaled based on local parameters and needs, or more preference could be given to indicators that are already localised and well understood by the community.

To use these tools and indicators, quality data needs to be collected from the SEPLS, which can be done using the following:

Citizen science: Through citizen science, people share their knowledge and contribute to data collection and monitoring, thus enhancing their participation and collaboration in scientific research (National Geographic, 2012). In the context of SEPLS, this could be seen as one of the most preferred methods for collecting local-level data, as it can directly engage the local community. It helps develop networks for ground monitoring, provided that it considers ecological knowledge, local use practices, tradition and heirlooms, and other socio-demographic heritage indicators.

Secondary data: Secondary data or outcomes can be examined, for example, from remote sensing data related to land use, mountainscape transformation, or climate change.

Established methodological frameworks and networks: Available frameworks and networks can provide data that help to provide basic ground information of the SEPLS. These include networks like long-term ecological monitoring networks.

2.3   What Are the Major Challenges in SEPLS M&E?

Despite the availability of different tools and data, practitioners face challenges in various contexts and levels. Some of the challenges are physical barriers and others are social or capacity barriers. A few of these are discussed below:

Developing linkages between various indicators and frameworks: Although various global indicators are available for sustainability, biodiversity, and health, they often suffer from various trade-offs among them. There could be a trade-off between the objectives of SEPLS and other objectives, such as the SDGs. For instance, the SDGs' "zero hunger" goal promotes free food distribution, affecting people's capacity and ability to grow local foods and develop local markets. This makes M&E a tough task from the nexus perspective, as while one goal is achieved, another is compromised. Therefore, there is a need to develop a global framework of indicators that allows for minimising trade-offs and maximising synergies to address the biodiversity-health-sustainability nexus and achieve multiple sustainability goals. Such a framework is currently lacking. Further, there is a lack of a system of global indicators that fully considers the nexus, which may lead to gaps between global indicators and local ones.

Lack of data and robustness: There is a lack of scientific data for M&E, and when available, they are often not scientifically robust. Quantitative data do not always address some socially relevant questions well, such as "why" a certain change happened. There is also a lack of knowledge on policy decisions that impact the lives of locals.

Monitoring after funding ends: There is often no monitoring after a project ends, as most donor-funded projects are output-focused and do not give much attention to long-term outcomes. For this reason, the actual effectiveness of the interventions is not measured over the long term.

Responsibility distribution for M&E: When multiple stakeholders are involved, the party responsible for M&E can remain unclear, and may depend on who owns the land, may it be private, government, or locals.

Quantitative nature of M&E: It is difficult to measure the benefits to a community, as they purely depend on what value community members put on the local resources, which broader quantitative estimates cannot measure.

2.4 What Are the Opportunities that Can Be Galvanised?

With challenges also come opportunities that can help overcome many barriers to effective M&E. These may include, but are not limited to:

Traditional knowledge and IPLCs: Local traditional knowledge and community-based monitoring can help with some of the listed challenges above, especially those pertaining to the lack of qualitative and on-the-ground data and responsibility for monitoring. Further, IPLCs also bring together years of accumulated knowledge and experience for better conservation and management. Thus, promotion of traditional knowledge in modern systems through participatory planning could be a way forward (e.g. Chaps. 2 and 3).

Mainstreaming of biodiversity and health: It is also worthwhile and relevant to mainstream health-biodiversity-sustainability indicators in other relevant indicators or approaches, to make them an integral part of ecosystem management and conservation (e.g. Chaps. 9 and 10).

Communication of M&E objectives and bridging stakeholders: In cases where the need for M&E and its implications to SEPLS management are clearly explained and well understood by the community and other stakeholders, the quality of the assessment process and its implementation can be enhanced (e.g. Chaps. 2 and 9). It is also desirable to bridge stakeholders including academic institutions, non-governmental organisations (NGOs), private companies, and other actors involved in facilitating SEPLS initiatives and fostering communication between primary (e.g. IPLCs) and secondary (e.g. government organisations) stakeholders in SEPLS. In many cases, these stakeholders play an essential role in conducting initial and follow-up M&E activities, in communication of M&E results and their management implications, and in capacity development of other SEPLS stakeholders (e.g. resilience assessment workshops in Xinshe SEPLS in Chap. 4).

3 How Can We Address the Challenges and Seize the Opportunities to Manage the Nexus in SEPLS?

To explore ways forward, the nexus approach helps to elucidate trade-offs associated with different pathways towards sustainability and to identify synergies that can build on the inherently coupled and mutually interdependent elements of SEPLS. This section first clarifies different types of trade-offs and then discusses the processes and steps to simultaneously attain multiple sustainability goals of SEPLS management through the nexus approach.

3.1 Types of Trade-Offs

Trade-offs are inevitable given the constraints in natural resources and multiple values attributed to them differently by different actors. The case studies show four types of trade-offs, including those between (1) ecosystem services, (2) stakeholders, (3) human well-being components, and (4) management goals. Due to the complex interlinkages between various elements of SEPLS, these types are not always clear-cut, but rather reciprocally influence each other. Furthermore, many of these trade-offs cut across sectors and scales, making varied impacts on different sectors, geographic and temporal scales, and governance levels. This typology is applicable to a wide range of multifunctional social-ecological systems (Lu et al., 2021; eds. Sarmiento & Frolich, 2020; Manning et al., 2018), but the case studies in this volume highlight the trade-offs in seeking for both ecosystem and human health and sustainability that can result from human-nature interactions within SEPLS.

Ecosystem Services

SEPLS, involving a variety of ecosystem functions and processes, provide multiple ecosystem services between which trade-offs can occur across sectors and scales. For instance, the case of the Angkorian landscape in Cambodia explicates the trade-offs between provisioning, regulating, and cultural services. The exploitation of forest resources (along with other drivers such as forest fire) has led to the reduction in water storage capacity and availability, affecting the quality and quantity of water in the upper catchment and threatening the downstream system of water management, which is also linked to agricultural production (e.g. through irrigation) and tourism (e.g. through flood control) (Chap. 7). In this case where tourism is highly dependent on natural and human-made water management networks, recent trends demonstrate that the excessively flourishing tourism (e.g. too many inbound tourists) contributes to the increase in demand for water that can put pressure on groundwater resources, not only affecting the water supply but also fragilising the structural base of the temples as cultural assets. More recently, the COVID-19 pandemic has adversely affected tourism (through decreased tourists) and has in turn reduced the revenue available for the water sector, constraining the water management administration despite the lessened pressure on water resources.

Trade-offs also arise between quantity and quality of the same kind of ecosystem services. The case of Indian peri-urban wetlands accounts for experiences in which curbing the use of chemical fertilisers and pesticides has improved the quality of food provisioning services and redressed negative health impacts on humans, but reduced the quantity of food provision in terms of crop production (Chap. 6). Similar trade-offs are found in the case of the Ghanaian cocoa farms, where government intervention to control cocoa pests and diseases has significantly increased the cocoa output nationwide but undermined the quality of cocoa production, imposing negative impacts on ecosystem and human health particularly for those within and around the local cocoa farms (Chap. 12).

Stakeholders

Given the varied ecosystem services derived from SEPLS, trade-offs arise likewise among the stakeholders who receive those ecosystem services that contribute to human well-being. The Angkorian case, where upstream forest degradation has lowered downstream groundwater levels, illustrates that the material gain by the population in the upper catchment (e.g. forest products) is traded off with the well-being of those in the lower catchment (e.g. access to groundwater) (Chap. 7). This raises a question about who should pay for the benefits from water management (including forest management), considering that efforts in forest conservation in the upper catchment, as well as water-efficient agriculture in the mid-catchment, would contribute to downstream water conservation.

These trade-offs also involve political dimensions in regard to the distribution of benefits from SEPLS management. The case of Alpine landscapes in Austria shows that ski tourism development has definitely benefited the international enterprises and their customers (e.g. through sports, wellness, and health), who can largely control the business with their monetary power, but has mixed effects on human and ecosystem health at the local level (Chap. 8). The regular tourism-based income has helped to stabilise the livelihoods of the locals and keep traditional sustainable practices alive (e.g. traditional gentian use contributing to biodiversity conservation) while limiting opportunities for alternative land uses (e.g. through the construction of large wellness hotels) and for earning decent subsistence from farming alone.

The case of local vegetable promotion in Africa also suggests that the gender gap in power raises trade-offs in regard to the benefits from the vegetable production: the women's contribution to growing the underutilised vegetables is not socially well recognised but instead, men take a central role particularly once the initiative becomes economically profitable (Chap. 10). This alludes to the gendered trade-offs in equitable sharing of benefits from the initiative (e.g. economic profits, social recognition). The same case also points to possible trade-offs between holders and users of traditional knowledge concerning local species and associated cultivation and utilisation practices. As the political and economic power of knowledge users mostly outweighs that of holders, particularly for traditional knowledge use, the rights of knowledge holders are often undermined without any regulations and their shared benefits from knowledge use could be compromised.

Human Well-Being Components

Related to the above trade-offs, different value preferences bring on trade-offs between human well-being components—including material, relational, and subjective dimensions of well-being. This kind of trade-off often happens between different stakeholders, and even within the same community and individual. The case study of the "ridge-to-reef" watershed in Chinese Taipei, for instance, signals the trade-offs incurred by the return of migrated youths to the community (Chap. 4). Some of the returned youths valuing quality of life surrounded by a wealth of nature (i.e. subjective well-being) became disillusioned with the rural living standard and fewer income-generating opportunities (i.e. material well-being) and went back to the city. Their continued urban lifestyle associated with wasteful consumption (i.e. material well-being of the returned youths) could be perceived by the locals as being against their cultural and traditional values (i.e. subjective well-being of the locals), leading to conflicts or undermining the social cohesion (i.e. relational well-being of both ends).

Another example is the Cambodian case of the Angkorian landscape where migration under the pandemic is leading to emerging trade-offs between food security and food sovereignty (Chap. 7). Food security of the locals (i.e. material well-being)—in addition to the security of land, water, and jobs—has declined due to the increased demand resulting from returned migrants in the aftermath of the COVID-19 outbreak. In combination with border closures and raised awareness of health, however, this challenge has turned their attention to local food and associated cultural practices, creating opportunities for food sovereignty through local food promotion and cultural preservation (i.e. subjective well-being). Nevertheless, achieving local sustainability may depend on whether the cost for local food production can be paid (i.e. material well-being) and whether the ownership of food, knowledge, and culture can be secured (i.e. subjective well-being). In this connection, the conflict between cocoa production and conservation in the Ghana case points to the importance of trade-off analysis that could evidence the ecological and health benefits from the integrated pest management (IPM) with biopesticide use compared to those from chemical pesticide use (i.e. material well-being) (Chap. 12). Such evidence helped increase the awareness among the farmers (i.e. subjective well-being), resulting in the increased adoption of IPM.

Management Goals

The sustainability goals of SEPLS management are indeed in multiples as similarly manifested in the SDGs. It is often the case that some goals could be compromised or even sacrificed to pursue other goals if it is not feasible to attain multiple goals at the same time. These trade-offs cut across economic, environmental, and social pillars of sustainability. For instance, conservation measures such as the mandatory ban of fishing during the spawning seasons of certain fish species (i.e. environmental) distress the livelihoods of those economically dependent on local natural resources such as traditional resource users (i.e. economic) (Chap. 2). The introduction of alternative livelihood means to them (e.g. aquaculture), however, could help to meet the original conservation goal (i.e. environment) without jeopardising their livelihoods (i.e. economic).

Even once a win-win situation is attained, economic betterment through natural resource use again requires caution for further trade-offs. The initiative to promote underutilised crops in Africa, for example, involves the threats of overharvesting and genetic erosion (i.e. environmental), particularly when it becomes economically profitable (i.e. economic) (Chap. 10). This threat could be furthered by attracting outsiders with different motivations or priorities, who may not only exploit natural resources for economic gain (i.e. environment vs. economic) but also undermine the social cohesion of the local communities (i.e. social). Similar trade-offs are suggested in the Indian case of peri-urban wetlands (Chap. 6). Some of those initially motivated in wetland conservation to develop ecotourism as an alternative livelihood activity could become further committed to tourism development (i.e. economic) but less attentive to conservation (i.e. environmental), especially once they are economically better off. Such a shift in mindset could divide stakeholders who were once more united to start up ecotourism with wetland conservation but started to seek different priorities and interests (i.e. social).

3.2 Ways Forward

To achieve multiple sustainability goals simultaneously, strategies for SEPLS management should essentially minimise inevitable trade-offs. Yet, solutions could be found not necessarily within the zero-sum game in resource allocation, but can build on synergies between multiple resources of SEPLS (Mohtar & Daher, 2017). The nexus approach helps address both trade-offs and synergies to seek for sustainability. Indeed, materialising the nexus approach in action allows us to pursue sustainable pathways. This process entails three steps: (1) identifying problems and solutions, (2) motivating multiple actors, and (3) governing multiple actions.

Identifying Problems and Solutions

To search pathways towards sustainability, it is highly critical to identify a "nexus hotspot". The hotspot is a vulnerable area of the "nexus" (i.e. wherein trade-offs are established) at a given scale, which faces stresses resulting from resource allocation that is at odds with harmonious human-nature relationships (Mohtar & Daher, 2016). Due to the multiscale nature of trade-offs, attention needs to be geared towards existing and potential nexus hotspots at multiple scales so as to minimise trade-offs and strengthen the weakest links to secure positive connectivity. This process requires attention to synergistic links between multiple resources so as to move beyond the zero-sum game for allocating fixed resources (Mohtar & Daher, 2017). The pathways could thus be better identified through dialogues and communications among multiple actors who value and bring together different resources and develop their own capacities for the nexus analysis (also see Chap 1).

Challenges and opportunities in drawing on the nexus analysis of problems and solutions lie in two levels: knowledge and information, and communications. First, knowledge and information concerning the nexus in SEPLS involve a high level of complexity, ambiguity, and uncertainties, giving rise to challenges. For instance, transition to a decarbonised society would require a major transformation, at least in tourism, energy, and water sectors in the Austrian context, but efforts to bring about such an unprecedented transformation may have unintended consequences, leaving equity-related questions about who will be benefited and who should be compensated for the costs for transition. Even in ongoing initiatives, challenges also exist. Adverse effects of pesticide use on human and ecosystem health and the productivity of cocoa farming have been felt and observed by local communities, and then their enhanced awareness has helped improve both ecosystem and human health synergistically at the local level (Chap. 12). Yet, these effects are not well recognised by policymakers, whereas fake or inaccurate information through media could mislead the stakeholders to take action. Also, transnational and multinational projects ignorant of local environmental conditions have made a huge investment in fossil fuel-based power plants in an ecologically critical area in Bangladesh.

Opportunities to overcome these challenges and forge synergies lie in the co-production of knowledge among multiple actors and making information more communicable to them (e.g. Sarmiento et al., 2013). Knowledge co-production among a variety of stakeholders—including local communities, government officials, enterprises, scientists, youths, the elderly, men, and women—allows for learning from their diverse experiences (including trial and error), bringing in new skills and different expertise, transferring knowledge within and across generations, and empowering stakeholders for their sound decisions and actions (e.g. Chap. 4). Among others, experiential knowledge held by local communities is essential for knowledge co-production in SEPLS, where trade-off impacts on ecosystem and human health are directly felt and seen. However, such knowledge is not always easily understandable and available in a communicable form to other stakeholders, including decision makers on the SEPLS management (e.g. policymakers). Some cases provide examples of efforts that have been made in this regard, such as the methodological development for documentation, synthesis, and visualisation of knowledge and information on foodways for better recognition and communication among different stakeholders including knowledge holders (e.g. Chap. 10).

Second, communication of knowledge and information is an important process for the nexus analysis. Bringing together a wide range of stakeholders in the communication process (e.g. the "whole-of-the-society" approach) to avoid negative consequences for a certain SEPLS segment, particularly those vulnerable to environmental change (e.g. IPLCs), and build synergies is indeed a high road. Yet, it is not always easy to create new communication channels between different groups. The case of a biocultural hotspot in India illustrates the distinctions between the four sectors—food, health, culture, and biodiversity—which have posed a challenge in communication and collaboration to promote the combined efforts on local health traditions and local food baskets, but which if integrated could offer opportunities for synergies (Chap. 9). The case of forest and milpa landscape (FML) in Mexico also shows the challenges in involving external actors with very different interests and values (e.g. pig firms and mega-projects) in the process of local SEPLS management (Chap. 5). This is even tougher in situations of conflict and uprising. The case of Sundarbans in Bangladesh alludes to the transboundary conflicts of water allocation, which heighten competition rather than cooperation for transborder watershed management (Chap. 2). In the case of the Colombian Pacific ecoregion, where some civil unrest has amounted to uprising and violence, even those who are committed to collaboration (e.g. scientists, NGO workers) can hardly reach out to local and indigenous communities for interventions in livelihood support and cultural preservation due to security issues (Chap. 3).

Even when all stakeholders cannot be brought in, it is worthwhile to create or join a platform for communication, dialogue, and learning so as to start converging the interests of different stakeholders. The FML case of Mexico illustrates the workshop-based peer learning through which multiple local communities within a wider landscape came to better understand broader perspectives on socio-economic conditions and collectively developed integrative bottom-up action plans for SEPLS sustainability (Chap. 5). Likewise, in the case of Chinese Taipei, regional efforts have in fact progressed to facilitate dialogue across a wider land/seascape, as part of a nationwide initiative, called the Taiwan Partnership for the Satoyama Initiative (TPSI), to identify common issues across different SEPLS communities and systematically develop locally sensitive management strategies. The TPSI and Taiwan Ecological Network—and the national policy that supports them—operate with three types of nexuses including the "green nexus" (terrestrial ecosystems), "blue nexus" (freshwater and marine ecosystems), and "human nexus" (multiple SEPLS stakeholders nationwide).[4] In addition, as suggested by the Colombian case, even in the case where local- or regional-level communications are difficult, global sharing

---

[4]For further information, please see (1) Taiwan Partnership for the Satoyama Initiative (TPSI), Forestry Bureau, Council of Agriculture official website (https://conservation.forest.gov.tw/EN/0002040) and (2) 2018–2021 Taiwan Ecological Network, Forestry Bureau, Council of Agriculture official website (https://www.forest.gov.tw/0002812) (in Chinese).

of information and experiences would offer opportunities to facilitate global knowledge co-production, replication of good practices elsewhere, and international collaboration to overcome challenges (Chap. 3).

Motivating Multiple Actors

The process of knowledge co-production through communication, dialogue, and learning helps motivate people to act on the findings from nexus analyses. It facilitates awareness raising, capacity development, and community empowerment so that stakeholders can better understand existing and potential trade-offs, recognise their rights and responsibilities, and appreciate their capacities and resources in managing SEPLS. This process can have positive effects particularly on the subjective and relational well-being of stakeholders as members of society moving towards sustainability, possibly leading to a shift in mindset and behavioural change from unsustainable to sustainable practices.

However, material well-being is no less important to motivate people to take action for sustainability. The case of Kampong Cham in Cambodia clearly shows that farmers' uptake and continuation of sustainable agricultural practices (e.g. composting and use of organic fertilisers) fundamentally rely on their cost-benefit analyses to sustain their livelihoods (Chap. 11). If additional costs are incurred through the introduction of alternative practices to livelihood means, such costs need to be sufficiently compensated; otherwise community members can hardly be motivated to engage in such practices. Indeed, economic incentives would work for garnering interest and getting key stakeholders engaged in new sustainability projects. The Indian biocultural hotspot case offers evidence that the premium price for medicinal rice with value addition, which was achieved through marketing development supported by the local government, helped to compensate for the crop loss caused by switching from the previous chemical use to organic farming (Chap. 9). This intervention has economically incentivised farmers and other stakeholders to support and engage in practices that synergistically enhance ecosystem and human health. As a result, it has been replicated beyond the state in the country.

Economic viability of interventions is needed for the long term, but the sustainable practices supported through subsidised projects often cease once the project terms end. Economic viability, however, cannot be built overnight. It is rather susceptible to market conditions and difficult to ensure over the long run. The Cambodian case of sustainable farming practices exhibits the challenge that even the markets, which were once viable for organic products, have become threatened by changes under the COVID-19 pandemic (e.g. the loss of access to major markets as well as the loss of tourists as consumers resulting from border closures, and increased demand for quantity of food due to the returned migrants) (Chap. 11). To make SEPLS more economically viable and more resilient to shocks and stresses, cautious approaches could be taken. One way is to put safety-net measures in place to limit the liability of negative trade-off impacts on local communities (e.g. a universal social security system to ensure access to health, education, and food security, microinsurance coverage for economic security) (e.g. Chaps. 2 and 6). Another is to build short-term measures within a SEPLS management programme (e.g. immediate income compensation, managerial and financial support) in consideration of a transition period for capacity development, which often takes time (e.g. Chaps. 5 and 6).

Governing Multiple Actions

As mentioned above, trade-offs occur across sectors, levels (governance), and scales (temporal and spatial). Governing different actions at smaller scales could be more feasible given that actors may identify trade-offs that are more visible at hand and could be motivated to develop synergistic solutions through more intimate communications and dialogues at the local level. However, managing the nexus that cuts across levels and scales requires coordination and implementation of activities at the large scale. This in turn provides opportunities to tap into diverse knowledge, expertise, and resources of a wider range of stakeholders while giving rise to challenges in mobilising and accommodating them and steering all of them towards sustainability. In this regard, governments fulfil an important role in enhancing policy coherence and ensuring fair and equitable incentive mechanisms across sectors and scales (Mohtar & Daher, 2017). The case of foodways documentation in Kenya alludes to the significance of policies, regulations, and protocols to prevent local natural resources from being exploited by outsiders and to protect the rights of traditional knowledge holders to be equitably benefited from knowledge use (Chap. 10).

Governments alone cannot manage the nexus. Their knowledge and information could be partial and might even be incorrect given the complexity, ambiguity, and uncertainties associated with the nexus, possibly misleading the governmental policies (e.g. Chap. 12). The "whole-of-the-government" approach may work to some extent to mobilise knowledge and resources from different sectors for more synergistic policymaking (e.g. cooperation between various agencies subordinate to the Council of Agriculture in Xinshe SEPLS, Chinese Taipei (Chap. 4)). Yet, knowledge and expertise at the government level can be insufficient to fully address existing and potential trade-offs and create synergies. Likewise, the media can serve as a major communication channel to cover and disseminate information, mobilise public opinion, and influence policies. Yet, their information could again be partial, biased, or wrong. If so, it may erroneously or even contradictorily influence policies, while the advancement of information and communication technology (e.g. social media) may offer opportunities to better inform policymaking and implementation.

Given that full-blown knowledge on the nexus cannot be produced instantly at a certain point of time and area, the governing process needs to be repetitive, cyclic, and gradual. Several case studies suggest that integrated and inclusive approaches to SEPLS management at the local level (e.g. the One Health approach) should be scaled up to address the issues of SEPLS that are interconnected and interdependent at higher levels and scales (e.g. biodiversity, climate change, livelihoods, pollution, gender, employment, and health) (e.g. Chaps. 2 and 9). As a first step to do so, it is again crucial to support and, if needed, strengthen community-based governance institutions to facilitate endogenous development and make the interventions more sustainable. For instance, the case from Colombia points to the significance of safeguarding the livelihoods and knowledge of indigenous peoples in reducing their vulnerabilities and leading to betterment of both human and ecosystem health through their enhanced ecosystem stewardship (Chap. 3). In this connection, the Indian case of peri-urban wetlands suggests that a phased approach should help to mainstream local institutions into higher level policy, comprising step-by-step interventions in SEPLS management (Chap. 6). The approach included introduction of microinsurance coverage to local communities, provision of alternative economic opportunities along with capacity development, consultation among different stake-holders with administrative and financial support, and awareness raising at the broader scale.

These governing processes can facilitate identifying and acting on synergistic solutions that build on local experiences and lessons but mobilise a broader range of knowledge, expertise, and resources to minimise trade-offs and create synergies. Synergistic examples at the relatively small scale include job creation based on traditional practices (e.g. food production and cultivation), which has led to sustain-able use of biodiversity, cultural preservation, land security, and income generation for the locales (e.g. Chaps. 2 and 7). Another example at the larger scale is the "ridge-to-reef" synergy where eco-agricultural practices in SEPLS have contributed to (1) improved health of the coral reef coastal ecosystem, (2) reintroduction of native species leading to revitalisation of indigenous and local knowledge, (3) crea-tion of new income-generating opportunities based on local natural resources and return of migrant youths, and (4) enhanced social capital of the communities allowing for more marketing opportunities and more resilient livelihoods (Chap. 4).

4 Conclusion

The interconnections between biodiversity, health, and sustainable development in the SEPLS context could be considered fundamental to ensure the well-being of the social-ecological system. A well-functioning and biodiverse ecosystem ensures attainment of various material benefits (e.g. food, medicine, clean water), regulating benefits of natural processes such as water cycles and nutrient cycles, and intangible and non-material benefits (e.g. cultural and educational values). These in turn provide a base for productive livelihoods and attainment of different components of good quality of life, which range across security of food, health, income, liveli-hoods, and identity, to name a few. Factors that enable achieving a good quality of life include access to biodiverse resources; knowledge related to their use, manage-ment, and value addition; adaptive capacity to economic and environmental shocks; and equitable and respectful interactions between different stakeholders in the governance and management of the SEPLS. That said, in practice, as the case studies highlight, several trade-offs arise owing to non-alignment or mismatches between sectoral goals, stakeholder priorities, and institutional mandates.

At the same time, opportunities to synergise resources can be leveraged across the phases of decision-making (i.e. planning, implementation, assessment, monitoring, and review) through participatory and inclusive approaches involving all stakeholder groups. The importance of peer learning, co-production of knowledge, promoting co-management and co-responsibility over resources and sectoral priorities, fostering bridging of institutions to negotiate and reduce conflicts, and engaging citizens to ensure "whole-of-society" commitment comes across quite clearly in this regard.

Given the high relevance of local actions to meeting national and global policy goals on sustainable development, evaluation, monitoring, and reporting on such activities to high-level planning forums are essential. This also needs to be done in a manner that ensures the richness of the contexts but speaks to the generalised language of policy forums. What is noteworthy, though, is that multiple synergies and trade-offs arising from the complex social-ecological connections in SEPLS challenge the monitoring of progress in the nexus context. This could be resolved through globally accepted indicators harmonised in the manner of minimising trade-offs and maximising synergies, and localised based on community needs and adaptability. It means that even while there exist no single methodology or standardised indicators to measure or assess the strength of the nexus between biodiversity, health, and broader sustainability goals, the case studies demonstrate that it is possible to do so by contextualising accepted indicators and engaging in more inclusive and participatory approaches to planning, managing, and governing SEPLS. However, these also require investments in data collection, mainstreaming activities, and capacity-building of different stakeholder groups to sensitively conceive of and implement SEPLS management plans taking into consideration multiple sectors and their overlapping agendas.

References

Bours, D., McGinn, C., & Pringle, P. (2014). Design, monitoring, and evaluation in a changing climate: Lessons learned from agriculture and food security programme evaluations in Asia, SEA Change and UKCIP.

GIZ, UNEP-WCMC, & FEBA. (2020). *Guidebook for monitoring and evaluating ecosystem-based adaptation interventions.* Deutsche Gesellschaft für Internationale Zusammenarbeit (GIZ) GmbH.

Hahn, M. B., Riederer, A. M., & Foster, S. O. (2009). The livelihood vulnerability index: A pragmatic approach to assessing risks from climate variability and change—A case study in Mozambique. *Global Environmental Change, 19*, 74–88. https://doi.org/10.1016/j.gloenvcha.2008.11.002

Lu, N., Liu, L., Yu, D., & Fu, B. (2021). Navigating trade-offs in the social-ecological systems. *Current Opinion in Environmental Sustainability, 48*, 77–84. https://doi.org/10.1016/j.cosust.2020.10.014

Manning, P., van der Plas, F., Soliveres, S., Allan, E., Maestre, F. T., Mace, G., Whittingham, M. J., & Fischer, M. (2018). Redefining ecosystem multifunctionality. *Nature Ecology & Evolution, 2,* 427–436. https://doi.org/10.1038/s41559-017-0461-7

Mohtar, R. H., & Daher, B. (2016). Water-energy-food nexus framework for facilitating multi-stakeholder dialogue. *Water International, 41*(5), 655–661. https://doi.org/10.1080/02508060.2016.1149759

Mohtar, R. H., & Daher, B. (2017). Beyond zero sum game allocations: Expanding resources potentials through reduced interdependencies and increased resource nexus synergies. *Current Opinion in Chemical Engineering, 18,* 84–89. https://doi.org/10.1016/j.coche.2017.09.002

National Geographic. (2012). Citizen science. *National Geographic Society,* viewed 22 August 2021. Retrieved from http://www.nationalgeographic.org/encyclopedia/citizen-science/.

Nishi, M., Subramanian, S. M., Gupta, H., Yoshino, M., Takahashi, Y., Miwa, K., & Takeda, T. (2021). Synthesis: Conception, approaches and strategies for transformative change. In M. Nishi, S. M. Subramanian, H. Gupta, M. Yoshino, Y. Takahashi, K. Miwa, & T. Takeda (Eds.), *Fostering transformative change for sustainability in the context of socio-ecological production landscapes and seascapes (SEPLS)* (pp. 229–249). Springer.

Sarmiento, F. O., & Frolich. L. M. (eds.) (2020). The Elgar companion to geography, transdisciplinarity and sustainability, Edward Elgar Publishing, 428pp. https://doi.org/10.4337/9781786430106.

Sarmiento, F. O., Russo, R., & Gordon, B. (2013). Tropical Mountains multifunctionality: Dendritic appropriation of rurality or Rhyzomic community resilience as food security panacea. In J. R. Pillarisetti, R. Lawrey, & A. Ahmad (Eds.), *Multifunctional agriculture, ecology and food security: International perspectives* (pp. 55–66). Nova Science Publishers.

UNU-IAS & IGES (Ed.). (2019). *Understanding the multiple values associated with sustainable use in socio-ecological production landscapes and seascapes (SEPLS)* (Satoyama initiative thematic review) (Vol. 5). United Nations University Institute of Advanced Studies in Sustainability.

WHO (Ed.). (2006). *Constitution of the World Health Organization* (45th ed.). WHO.

The opinions expressed in this chapter are those of the author(s) and do not necessarily reflect the views of UNU-IAS, its Board of Directors, or the countries they represent.

Open Access This chapter is licenced under the terms of the Creative Commons Attribution 3.0 IGO Licence (http://creativecommons.org/licenses/by/3.0/igo/), which permits use, sharing, adaptation, distribution and reproduction in any medium or format, as long as you give appropriate credit to UNU-IAS, provide a link to the Creative Commons licence and indicate if changes were made.

The use of the UNU-IAS name and logo, shall be subject to a separate written licence agreement between UNU-IAS and the user and is not authorised as part of this CC BY 3.0 IGO licence. Note that the link provided above includes additional terms and conditions of the licence.

The images or other third party material in this chapter are included in the chapter's Creative Commons licence, unless indicated otherwise in a credit line to the material. If material is not included in the chapter's Creative Commons licence and your intended use is not permitted by statutory regulation or exceeds the permitted use, you will need to obtain permission directly from the copyright holder.